Creativity and Innovation in Business and Beyond

RIOT! Routledge Studies in Innovation, Organization and Technology

Creativity and Innovation in Business and Beyond

Social Science Perspectives
and Policy Implications

Edited by Leon Mann and Janet Chan

Routledge
Taylor & Francis Group
New York London

First published 2011
by Routledge
711 Third Avenue, New York, NY 10017

Simultaneously published in the UK
by Routledge
2 Park Square, Milton Park, Abingdon, Oxon OX14 4RN

Routledge is an imprint of the Taylor & Francis Group, an informa business

First published in paperback 2012

Typeset in Sabon by IBT Global.

Library of Congress Cataloging-in-Publication Data
    Creativity and innovation in business and beyond : social science perspectives and policy implications / edited by Leon Mann and Janet Chan.—1st ed.
        p. cm.—(Routledge studies in innovation, organization and technology ; 18)
    Includes bibliographical references and index.
    1.   Business enterprises—Technological innovations. 2.   Technological innovations—Management. I. Mann, Leon. II. Chan, Janet B. L.
    HD45.C6884 2011
    338'.064—dc22
    2010028965

ISBN13: 978-0-415-88010-7 (hbk)
ISBN13: 978-0-203-83306-3 (ebk)
ISBN13: 978-0-415-64898-1 (pbk)

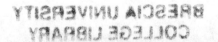

For Michelle and Lisa and
for Karen and Kenneth
May their lives be creative and fulfilling

Contents

Tables

Figures

Acknowledgments

This research was partly funded by an Australian Research Council Linkage Academy Special Project Grant through the Academy of Social Sciences in Australia (ASSA). We would like to thank John Robertson, the former ASSA research director, for his contributions to the early stages of the project. The project also benefited enormously from the ideas and discussion during the ASSA symposium "Fostering Creativity and Innovation" that we organised in Canberra in November 2008. The symposium pointed us to other disciplines with a stake and an interest in creativity and innovation, highlighted the relevance of these ideas for higher education and other policies and reminded us that social science should be concerned with sometimes negative disruptive impacts of innovations.

We would also like to thank Laura Stearns, the commissioning editor of Routledge, for recognising the significance and distinctiveness of our collection. The production of this book benefited from the valuable assistance of Stacy Noto and Roanna Gonsalves. Their work is greatly appreciated.

Preface

The title of this book, *Creativity and Innovation in Business and Beyond*, indicates the scope of this volume: creativity and innovation can be found in every area of human enterprise—in art and design, music and literature, science and technology, research and development. They can benefit every aspect of human organisation: from medicine and health, to education and training, security and safety, housing and transport, information and communication, commerce and trade, recreation and entertainment. The capacity for creativity and innovation has enabled us to evolve, adapt and survive as humans. It has underpinned our march toward civilisation, from harnessing fire, to shaping the first tools, to inventing the wheel, to organising for hunting, to constructing canoes and rafts, to making glass, to farming and harvesting. It has also underpinned modes of expression and communication from the very beginnings in rock carving and cave painting, to making and decorating vessels, to ceremony and dance, to writing and record keeping, to the printing press and beyond. Yet the drivers of creativity and innovation are not well understood and historically have been researched as separate topics. This book is a first attempt to examine the relationship between the two processes.

The subtitle *Social Science Perspectives and Policy Implications* reflects the distinctiveness of this volume in bringing together the insights of a group of internationally renowned social scientists from a diverse range of disciplines. Disciplines such as psychology, sociology, education, cultural anthropology and social history have had a long history of studying *creativity* as a human process and achievement, while disciplines such as economics, including the sub-disciplines of economic geography and economic history, law and management, have made significant contributions to the understanding of *innovation* in organisational and technological advancement. Together these disciplines provide a range of perspectives on the meanings, sources and drivers of creativity and innovation. They invite readers to think about the connection between creativity and innovation, and suggest how governments and organisations can provide the most favourable environment to engender a culture of creativity and innovation.

Research and writing are themselves creative endeavours. We are delighted to have been part of a challenging research and writing project with an outstanding team of scholars to bring together some key ideas and questions at the core of creativity and innovation.

Leon Mann and Janet Chan
Melbourne and Sydney
June 2010

1 Introduction

Creativity and Innovation

Janet Chan and Leon Mann

WORLDWIDE INTEREST IN CREATIVITY AND INNOVATION

The terms "creativity" and "innovation" traditionally referred to separate spheres of social life: creativity was primarily associated with activities of the arts, while innovation was linked to scientific discoveries or technological advances. Where creativity evoked images of paintings and sculptures, innovation brought to mind new products and technologies. This apparent separation was further exacerbated by the link between technological innovation and economic productivity, while the creative arts were considered socially valuable but nonetheless "non-productive labour" (Potts 2007). Hence the creative arts were sustained by government welfare in the form of grants, subsidies or special funds, whereas science and technology received both public funding and private investments.

The situation changed dramatically around the beginning of this century. "Creativity" is now increasingly twinned with "innovation" and mentioned in association with the economy. Here are some examples. A 2008 government report *Creative Britain: New Talents for the New Economy* saw "creativity as the engine of economic growth for towns, cities and regions" (UK Department for Culture, Media and Sports [DCMS] 2008, 6). It called for public funding to "stimulate creativity and sharpen Britain's creative edge" and set out a blueprint to make Britain "the world's creative hub" (2008, 6). The European Union declared 2009 the European Year of Creativity and Innovation.[1] The aim was to "raise awareness of the importance of creativity and innovation for personal, social and economic development; to disseminate good practices; to stimulate education and research; to promote policy debate on related issues" (European Union 2009). In Australia, a report, *Imagine Australia*, prepared for the Prime Minister's Science Engineering and Innovation Council (PMSEIC), argued strongly for a "comprehensive approach to fostering creativity" in order for Australia to be "globally competitive" (Australia, PMSEIC 2005, 5).

Why is the term "creativity" now so closely linked to innovation and the economy? One obvious reason is that over the past decade, the notion of a "creative economy" has come to the forefront in public policy in many countries. The decline of the industrial economy has highlighted the importance of the so-called "knowledge economy" and, more recently, the "Creative Class" (Florida 2003) as a driving force of this new economy. The report *Creative Britain* states that in the UK, the creative sector employs two million people, contributes £60 billion a year to the British economy and has grown at double the rate of the economy as a whole over the last ten years (UK DCMS 2008). The report defines the creative sector as including "advertising, architecture, the art and antiques market, crafts, design, designer fashion, film, interactive leisure software, music, the performing arts, publishing, software and computer services, television and radio" (2008, n. 1).

The new interest in creativity as a partner to innovation is not limited to the creative industries; the idea is that creativity should be applicable to all areas of the economy (Redhead 2004). The *Imagine Australia* report explicitly discusses "the role of creativity in the innovation economy". It is in favour of investing in creativity so as to develop a "creative and innovative society, culture and workforce" (PMSEIC 2005, 7). The report includes a quote from the prime minister of Singapore: "For many years we concentrated on the economic side. But if you want the economic side to flourish, you need more entrepreneurs, you need more creativity. The two must go together" (2005, 6). The European Union similarly advocated nurturing creativity "in a lifelong learning process where theory and practice go hand in hand"; making schools and universities "places where students and teachers engage in creative thinking and learning by doing"; transforming workplaces into "learning sites"; promoting a "strong, independent and diverse cultural sector", scientific research and "design processes, thinking and tools"; and supporting business innovation that "contributes to prosperity and sustainability" (EU 2009).

In the world of high-tech industry, the processes of creativity and innovation are viewed increasingly as highly connected as some of the world's leading companies search for new ideas in the areas of pharmaceuticals, electronics, optics, information technology, energy, etc., for the "engines of tomorrow" (Buderi 2000): the more original the ideas generated for new products and services, the more successful the companies will become. Large innovative multinational companies such as Proctor & Gamble, Eli Lilly and IBM look to the "ideas pipeline" for new ideas sourced from leading-edge customers, firms in other industries, university laboratories and their own research centres. Companies such as Rolls-Royce, Pfizer and Microsoft also establish research collaborations and knowledge networks to access novel ideas and new knowledge external to the organisation for translation of that knowledge into company innovation. The quest by high-tech industry to find or discover new, creative ideas to turn into new products and services has spurred a growing interest in how best to manage creative researchers and professionals (see Mann, Chapter 10 this volume), how to foster and develop creative ideas and concepts in organisations (see

Dodgson, this volume) and how to design research organisations to balance creativity and control (West, Chapter 12 this volume).

Unfortunately some of these new developments linking creativity and innovation have been accompanied by hype and exaggeration, especially in the tendency by some commentators to uncritically describe any new idea as creativity and any change as innovation. As one author observes:

> creativity and the creative industries have been oversold. Definitions of creativity and the creative industries have been deliberately extended and manipulated, partly for self-serving reasons, partly to paint an inviting picture of our social and economic futures. In the process, the idea of creativity has been disconnected from the values which give it meaning and reduced to a banal pursuit of novelty. The creative industries have been similarly decontextualized, singled out for special mention as the cutting edge of a new economy. In reality the creative industries and the creative economy represent a shift in the way the economy as a whole is functioning, rather than a coherent category or industry sub-sector. (Bilton 2007, xx)

Similarly, Richard Florida warns against seeing creativity as a commodity: "Creativity comes from people. And while people can be hired and fired, their creative capacity cannot be bought and sold, or turned on and off at will" (2003, 5).

There are other good reasons to be cautious about jumping on the creativity–innovation bandwagon. Words such as *creativity* and *innovation* are "in danger of being over-worked and over-hyped" (Cutler Report 2008, 15). Like many other buzzwords such as community, partnership, social capital, etc., creativity and innovation are "hurrah" words that are tagged on to policies or programs to make them seem new and ground-breaking. Creativity, according to a critic, is a word that has been "debased" by "a generation of bureaucrats, civil servants, managers and politicians, lazily used as political margarine to spread approvingly and inclusively over any activity with a non-material element to it" (Tusa 2003, 6). Bruce Nussbaum went as far as pronouncing that "innovation" is dead: "It was done in by CEOs, consultants, marketers, advertisers and business journalists who degraded and devalued the idea by conflating it with change, technology, design, globalization, trendiness, and anything 'new'" (Nussbaum 2008).

MYTHS AND MISCONCEPTIONS ABOUT CREATIVITY AND INNOVATION

The hype about creativity and innovation is also exacerbated by a host of cultural myths and misconceptions that are not based on serious or systematic research. Berkun (2007) made a list of ten common myths about innovation, some of which also apply to creativity (cf. Sawyer 2006). For example, there is a myth that creativity is about personal talent. According to this myth, creative people are either geniuses or slightly mad. In other words, creativity is either a gift that a person is born with or a consequence of some mental abnormality. It is exemplified by the fairy tales of overnight success when a star is born through the discovery of hidden talent. Such a view ignores the knowledge, skills, training and persistent work required in creative achievements. It also underplays the degree to which creativity is the result of mutual help and collaboration between people.

A second myth cited by Berkun (2007) is that of the "lone inventor". As pointed out by Robert Merton (1961) and by Dean Simonton (2003), scientific "multiples"—multiple discoveries made by two or more scientists independently—are actually quite common: Simonton's (2003) research found 449 doublets, 104 triplets, eighteen quadruplets, seven quintuplets and one octuplet. The common occurrence of multiples can be explained by the way science works: the accumulation of scientific knowledge, the frequent interaction between scientists and the institutionalisation of scientific methods, so that "[o]nce the needed antecedent conditions obtain, discoveries are offshoots of their time, rather than turning up altogether at random" (Merton 1961, 473). However, Simonton's analysis suggests that scientific creativity exhibits the characteristics of a constrained stochastic process: it "demands the intrusion of a restricted amount of chance, randomness or unpredictability" (2003, 476). This means that, given the way scientists work, the probability of several researchers independently making the same discovery increases with the number of scientists working on the same research question, using the same techniques and concepts.

Another myth is that innovation involves "Eureka" moments—discoveries and breakthroughs that are *dramatic* or even earth shattering. As Eureka moments are rare in the history of science, breakthroughs are seen as a matter of luck or accident, rather than persistent hard work. In fact, discoveries and innovations involve many incremental and plodding steps that precede the final breakthrough. As Berkun points out, "there is no singular magic moment; instead, there are many smaller insights accumulated over time" (2007, 7): nearly all innovations of the twentieth century (the Internet, the Web browser, the search engine, and so on) are the result of years and even decades of work and the contributions of numerous people and organisations.

Finally, there is a misconception that creativity is a universal quality independent of social and historical context. The misconception is based on the assumption that the novelty and value of a creative product or innovation will be recognised and acknowledged whenever and wherever it occurs. Yet the judgment of a product as creative, as "new and valuable", as "original" and so on, changes with time and depends on a host of factors. An artwork may be regarded as ordinary or brilliant as determined by the current sensibility of the art world (Danto 1964). Similarly, a scientific breakthrough may not be recognised as such until there is a "paradigm shift" in the way the scientific community looks at a problem (Kuhn 1962/1970). "Novelty", as Bilton (2007, 3) points out, "is always relative". A creative product is judged by the relevant authorities in a *field*—for example, artists, art teachers, critics, collectors and curators—who operate within a *domain* of knowledge (Feldman, Csikszentmihalyi and Gardner 1994). If an artwork or innovation is too far ahead of its time, it may be seen as novel without being valuable (Bilton 2007). Creative products that break boundaries or are too far "outside the box" are in danger of being ignored or dismissed.

An aim of the present volume is to debunk common myths and misconceptions about creativity and innovation by clarifying the meanings of and the connections between these terms. We do this by reviewing the social science research literature and drawing on the research undertaken by the volume's contributors to identify the conditions for fostering creativity and innovation.[2]

THE MEANINGS OF CREATIVITY AND INNOVATION

In everyday usage the adjectives "creative" and "innovative" are more or less interchangeable: being creative is no different from being innovative. Thus, we can talk about a creative (or innovative) person, an innovative (or creative) solution to a problem or a creative (or innovative) product—the crucial feature is that the solution or product is new, novel or original. But the abstract noun "creativity" is not the same as "innovation". Very often, creativity refers to the quality of being innovative in thinking, planning or doing, whereas innovation refers to the end result of such creative thinking, planning or doing. Creativity is also conceived as a capability or a pattern of behaviour. Thus, the *Imagine Australia* Report sees creativity as "an innate and universal human trait", an "imaginative capacity to generate new ideas, images and ways of thinking; new patterns of behaviour; new combinations of action" (Australia, PMSEIC 2005, 6). In contrast, innovation is a product, a process or a solution that is new, revolutionary or inventive. Cutler defines innovation as "creating value through doing something in a novel way", or simply "good ideas put to work" (Cutler Report 2008, 15). Pratt and Jeffcut summarise some of the ways a distinction between the two concepts is made:

> Certainly, it is common knowledge that creativity is the "ideas" part of innovation; innovation usually being characterized as the practice of implementing an idea. . . . Others dispensed with creativity altogether replacing it with stages of innovation . . . For others still, creativity is quite different from innovation. Creativity encompasses new knowledge, whereas innovation may not be creative and can be incremental . . . Despite their differences most points of view acknowledge that context is important for innovation and creativity. (2009, 4)

They suggest that creativity and innovation should be "addressed as a *process* (requiring knowledge, networks, and technologies) that enables the generation and translation of novel ideas into innovative goods and services" (2009, 4). It is important to recognise, however, that the meanings of these concepts are not static—creativity and innovation are increasingly defined in terms of "the social and collective" rather than the individual as a result of social, cultural and technological changes (Hall 2010, x).

The Creativity Continuum

Creativity research within psychology has been divided into two camps: those interested in "Big-C" creativity and those more interested in "little-c" creativity. Big-C (eminent) creativity refers to creative works and creative genius, while little-c (everyday) creativity focuses on problem-solving activities that non-experts engage in routinely. Examples of Big-C creativity are the works of geniuses and masters (cf. Simonton 1994, 2004). As explained by Dean Simonton in an interview on creativity and intelligence:

> when you are talking about Big-C creativity . . . you're talking about being able to generate new ideas, generate some kind of product that's going to have some kind of impression on other people. It may be a poem, it may be a patent, it may be a short story, it may be a journal article or whatever. But it's something that is a concrete, discrete product that is original and serves some kind of adaptive function. (Plucker 2003)

Big-C involves solutions to very difficult problems and the ultimate Big-C is the production of significant original works. Big-C is the stuff of deep domain knowledge, schematic thinking, substantial memory and so on (see Sweller and Mann, this volume). In some circumstances Big C-creativity leads to the transformation of a domain.

In contrast, little-c creativity refers to simple problem-solving activities people engage in every day, e.g. modifying a recipe because an ingredient is missing, figuring out a short cut to avoid a traffic jam, using a coat-hanger to make a tool for retrieving an object. We make use of little-c creativity all the time. That basic problem-solving kind of creativity is very closely related to intelligence. The study of little-c is based on the assumption that creative potential is widely distributed and that all people can be creative (Kaufman and Baer 2006; Sternberg, Grigorenko and Singer 2004). Little-c is democratic and domain general. It is encouraged in schools and training programs. It is relevant to everyday innovation in how people work, for example, in changing practices in the hospitality and services industries and how people make clever use of new technologies such as the Internet and the iPhone.

David Feldman (see Morelock and Feldman 1999) reconceptualised the two opposites into a creativity continuum. He saw a place for middle-level creativity in the range between Big-C and little-c. Middle-c creativity refers to creativity in the expression of professional expertise. It is about creative products appreciated for interpretive skill, mastery of technical terms, distinctive style and success in achieving a technical, practical, commercial or academic goal.

In this volume we are primarily interested in middle-c and Big-C creativity inasmuch as we have an interest in creativity that leads to new and improved products and services and the underpinnings of radical, transformative innovation.

Bringing Creativity and Innovation Together

While creativity and innovation are bracketed in popular and management writing, there has been little attempt to bring them together in a systematic conceptual and empirical analysis. The connection is assumed but seldom elaborated. Mark Dodgson (this volume) argues that there is "a profound disconnect" between the treatment of creativity and innovation in the management literature, as evidenced by their separation as concepts in journal articles. Recently the social science disciplines have begun to join together in studying creativity and innovation, for example, in tracing the links between creative media, technology and industry; in mapping hubs of creativity and innovation in cities and regions such as Boston and Route 128, San Francisco and Silicon Valley; and in examining creative individuals and teams in the context of organisational innovation.

Many pathways connect creativity and innovation. Sometimes they connect as a sequence of milestones on a journey from creativity to innovation; sometimes they are a network of criss-crossing roads in which creativity and innovation intersect and interact across different domains; sometimes they are parallel paths heading in the same direction inspired by a larger event or necessity. The two processes are closely related, but are not identical. Some writers, from a creativity perspective, view innovation as an outcome of creative activity. Other writers, from an innovation perspective, treat creativity as a subset of innovation (see Mann, Chapter 15 this volume).

SOCIAL SCIENCE DISCIPLINES . . . AND SOME ABSENTEES

Some social science disciplines say more about creativity while others say more about innovation. For example, psychology and sociology tend to focus on creativity. The individual's capacity to create, whether alone, in teams, in organisations or as part of a community or social network, is an obvious starting point for analysis. In this volume we see how creativity is a function of the individual's cognitive architecture—thinking, information-processing and memory processes (Sweller and Mann), of teams and communities of practice (Mann, Chan), of organisational "play" (Dodgson) and of education in schools (McWilliam). In contrast, economics, history, political science, law and to some extent sociology tend to focus on innovation. Social, economic and historical forces are the natural starting points for an analysis of innovation. In this volume, we see how innovation is nested within national innovation systems and institutional frameworks (West, Marceau, Gans, Christie, Fitzgerald), operates differently across industry sectors (Marceau), takes place in cities and regions (Marceau), is grounded in historical and cultural forces and transformations (Ville) and is fostered and managed through research organisations, public and private (Dodgson, West, Chan, Mann).

The special contribution of the social sciences to the systematic study of creativity and innovation is that the two processes are examined as distinctive subjects based on theories, methods and tools of the social sciences. While all social science disciplines do not take the same approach to questions about creativity and innovation, they hold the promise of greater integration through multilevel analysis to provide an understanding of the critical factors that are the drivers at each level. The ideas and evidence presented in this volume are based on the research methods and approaches of the social sciences. Because of their diverse subject matter—from revolutionary transitions in the history of innovation to components of individual creative capacity—social science researchers tend to use a great variety of research methods. These methods include case studies, longitudinal studies, field and laboratory experiments, interviews, controlled observation, program interventions, surveys and panel studies, archival analysis and so on. Several social science disciplines that contribute to knowledge about creativity and innovation are absent from the volume. They include cultural anthropology, demography and political science. Several sub-disciplines of the social sciences are also absent. Economic history is included but not social and cultural history; management of innovation is included but not entrepreneurship; social psychology and cognitive psychology are included but not lifespan and developmental psychology. Also absent from the volume are several disciplines outside the social sciences that contribute to the study of creativity and innovation; a prime example is the history and philosophy of science. The reason for the absent disciplines and sub-disciplines is the near impossibility of covering all disciplines in a volume that is not a handbook or encyclopaedia.

SCOPE AND FOCUS OF THE CHAPTERS

The editors asked chapter writers to do three things: (a) define from the perspective of their discipline what is meant by creativity and innovation and how they are related, (b) describe some of their own research work and findings in the area and, finally (c) put forward recommendations for policy and practice based upon their analysis. Not unexpectedly, the authors tended to concentrate on either creativity or innovation, depending on their disciplinary orientation and research area.

Jonathan West's "National Innovation Systems and Creativity" reviews research into national systems of innovation: the context within which innovation and creativity occurs. The research finds that patterns of organisation for innovation vary considerably from nation to nation and that superior performance stems from finding an appropriate fit between the national social system—its inheritance of culture and institutions—and organisational practice. A detailed example drawn from the author's own research in the global semiconductor industry is provided as an example of several dimensions of this issue. The chapter concludes with consideration of the implications of these findings for the development of effective governmental policy to promote innovation.

In "Innovation and Creativity in Industry and the Service Sectors", Jane Marceau presents different models of the innovation process and the potential links between innovation and creativity. She suggests that innovation takes place in a framework shaped by the nature and interaction of eight different "ingredients"—human capital, science systems, research and technology uptake, technological change, intellectual property regimes, business services, trade and venture capital. Innovation is then determined by dominant factors in different industries, in some by knowledge flows between users and producers, in others by power relations between key players. This approach is illustrated by analysis of the drivers of innovation in a quintessentially "creative" industry—textiles and clothing.

In "Space, Place and Innovation", Marceau takes an economic geography perspective to examine the patterns and stimulants of innovation that are found in spaces and places. She explores innovation drivers in different spaces. After presenting current ideas on national and sub-national systems of innovation, the chapter examines differing views on what makes some cities innovation hubs. Is it the density of population which makes ideas and new knowledge flow relatively freely? Is it the "people equation", the type of people attracted to certain cities and the amenities and "atmosphere" they offer and the skills they bring with them? Or the better governance structures some cities have developed? The chapter finishes with a brief discussion of some technological contributions to urban life which may involve the whole population of a city and its government in doing many things in new ways.

Simon Ville's "Historical Approaches to Creativity and Innovation" traces the sweep of technological changes that underpin major innovation in key sectors of an economy. He explains that historians have focused more on innovation than creativity since it is outcome-based and easier to substantiate through historical evidence. A comparative analysis of the role of innovation in economic development and performance across time is provided in the chapter, which addresses whether the spreading effects of "macro-innovations" or a societal propensity to innovate has been a more powerful force for change, along with the methods of technology transfer and the significance of being an originator, user or adapter of technology. Australia's experience demonstrates the innovativeness of resource-based economies and the role of innovation as a driver of structural change.

In "Economic Approaches to Understanding and Promoting Innovation" Joshua Gans examines the economics of innovation and creativity. It starts with the notion that from an economic perspective the issue is how to allocate scarce resources (skilled scientists, R&D capital and the like) to innovative activity but that there are several constraints on the operation of markets that suggest that such allocation will not be socially optimal. These constraints provide a guidepost for policy interventions and collective institutions. The chapter reviews these with the goal of identifying when each might be most effective.

In "Creativity and Innovation: A Legal Perspective" Andrew Christie points out that in law, creativity and innovation are alternative thresholds of intellectual ingenuity that must be satisfied for intellectual property matter to gain protection. Where an intellectual property subject matter results from the exercise of creativity, the law may provide the protection of a copying right. Where, however, the subject matter results from a higher degree of intellectual ingenuity, such that it amounts to an innovation, the law may provide the stronger protection of monopoly rights. The chapter explains this outcome, the policy consequences of this outcome and why the policy consequences of this outcome are justifiable.

Brian Fitzgerald's "Promoting Creativity and Innovation through Law" considers the role of law in facilitating creativity and thereby sponsoring innovation. It argues that the exchange and reuse of knowledge and culture are vital ingredients of creativity and innovation. In turn it suggests that in order to facilitate creativity and innovation the development and interpretation of the law needs to promote the flow (exchange and reuse) of knowledge and culture. The chapter provides examples of how the law currently hinders information flow and makes recommendations as to how this could be remedied.

In "Towards a Sociology of Creativity" Janet Chan explores the elements that can make up a sociology of creativity. The chapter begins by examining the definitions of creativity and reviewing the literature on creativity to identify the social dimensions of creativity. It goes on to analyse Hans Joas's theory of creative action and Pierre Bourdieu's theory of practice to arrive at a sociological framework for understanding creativity. The framework provides a coherent and useful way of conceptualising and explaining creative practice as a product of the interaction between institutionalised structures and human agency.

In "Social Psychology of Creativity and Innovation" Leon Mann discusses the contribution of a social psychological approach for understanding the drivers of creativity and innovation in research teams. Drawing on studies of research teams in Australian research organisations, he shows how team climate and other social processes, such as leadership roles and style, trust, tacit knowledge and knowledge building and sharing, are significant for team creativity and innovation in all teams but especially for those engaged in fundamental research. Research teams are nested within organisations. The chapter shows how the organisation's climate on three dimensions—encouragement of innovation, provision of resources and empowerment—also affects the innovative performance of teams.

Mark Dodgson's "Creativity and Innovation Management: Play's the Thing" explores the relationship between innovation and creativity in the management literature, and finds a wide disconnection between them in the current literature. The association between innovation and creativity is argued to be important for business and management, but a novel empirical analysis shows scholars are concerned with one or the other, and rarely both. The chapter proposes that the concept of play provides a means of integrating the management of innovation and creativity, especially when allied to the use of new technologies such as virtual reality. Four short case studies are presented to illustrate the argument.

In "Inducing and Disciplining Creativity in Organisations under Escalating Complexity" Jonathan West explores the problems inherent in organising creativity and innovation. He argues that the organisational structures and procedures developed by managers of large organisations to pursue routine economic activity inevitably inhibit creativity and innovation. It takes an information-processing view of organisation to show that mounting complexity and uncertainty induce organisations either to centralise or to decentralise decision-making, and that the organisational forms induced by these paths create divergent challenges for innovators. The chapter offers perspective on approaches both innovators and managers of organisations that seek to promote innovation can employ to respond to these challenges.

In "Creativity and Innovation: An Educational Perspective" Erica McWilliam elaborates the contribution of education to creativity and innovation by exploring two centuries of educational theorising and practice, to consider what they bring to twenty-first-century understandings of creativity and innovation. The chapter begins by attending to definitional issues and historical background, before moving to consider substantive contributions of education under the following headings: "Creativity and Education of Different Kinds", "Creativity and Purpose: Humanistic versus Utilitarian", "Creativity and Education at Different Levels" and "Creativity and Individual versus Social Process". It concludes with a consideration of some policy implications for contemporary and future education.

In "The Psychology of Creativity and Its Educational Consequences" John Sweller and Leon Mann take an evolutionary approach to human cognition and consider the consequences of cognitive architecture for becoming creative in a particular field or domain. It is concluded that while creativity can be learned and developed through deep immersion and experience in a particular domain, there is little reliable evidence based on properly controlled experiments that it can be fostered through training and instruction in domain-general strategies. This has implications for how psychologists understand creativity and for education practice and training.

In "Creativity Meets Innovation: Examining Relationships and Pathways" Leon Mann identifies a variety of relationships between the two, including *causal*, in which creativity spurs innovation or technological innovation triggers a surge of creativity, *systems-dynamic*, in which creativity and innovation reinforce each other in cyclical feedback loops, and *correlational*, in which creativity and innovation overlap through a common set of underlying factors. Mann calls for the systematic study of the relationship between creativity and innovation to provide a foundation for understanding and fostering the drivers of both processes.

In the final chapter, "Creativity and Innovation: Principles and Policy Implications", Leon Mann revisits key themes in the book—that creativity and innovation are dynamically interconnected across a wide range of domains encompassing the arts, design, science and technology; that creativity and innovation take place in a range of settings—regions, cities, institutions, knowledge communities, organisations and teams; that while complex, it is possible to nurture and enhance creativity and innovation; that myths about creativity and innovation that hinder understanding must be dispelled; that a variety of research methods and approaches from the social sciences can be applied in the quest for multidisciplinary integration in the study of creativity and innovation. The chapter concludes with a section on recommendations for policy and practice for creativity and innovation across a number of levels, from national innovation systems, to industry sectors, cities and regions, organisations, schools and government, and legal, regulatory and economic instruments and incentives.

NOTES

1. The EU was explicit in its promotion of both creativity and innovation, nominating two particular competencies from the European Commission Lifelong Learning Framework: "Sense of initiative and entrepreneurship", defined as "an individual's ability to turn ideas into action . . . [including] creativity, innovation and risk-taking" and "Cultural awareness and expression", defined as an "appreciation of the importance of the creative expression of ideas, experiences and emotions in a range of media" (EU 2007, 11–12).
2. Another unfortunate myth one of us (LM) encountered while a member of the Australian government panel from 2002 to 2003 recommending national research priorities was the opinion from a government bureaucrat that

creativity is not a researchable topic. Fortunately Leon won the argument. Since 2003 "Promoting an innovation culture and economy: maximising Australia's creative and technological capability by understanding the factors conducive to innovation and its acceptance" has been designated a National Research Priority Goal in Australia.

BIBLIOGRAPHY

Amabile, T. M. 1996. *Creativity in context: Update to the Social Psychology of Creativity*. Boulder, CO: Westview Press.
Australia, Prime Minister's Science Engineering and Innovation Council. 2005. *Imagine Australia: The role of creativity in the innovation economy*. http://www.dest.gov.au/NR/rdonlyres/B1EF82EF-08D5-427E-B7E4-69D41C61D495/8625/finalPMSEICReport_WEBversion.pdf.
Berkun, S. 2007. *The myths of innovation*. Sebastopol, CA: O'Reilly Media.
Bilton, C. 2007. *Management and creativity: From creative industries to creative management*. Carlton, VIC: Blackwell.
Buderi, R. 2000. *Engines of tomorrow: How the world's best companies are using their research labs to win the future*. New York: Simon and Schuster.
Cutler Report. 2008. *Venturous Australia: Building strength in innovation*. Report of the Review of National Innovation System. http://www.innovation.gov.au/innovationreview/Pages/home.aspx.
Danto, A. 1964. The artworld. *Journal of Philosophy* 61: 571–84.
European Commission. 2007. *Key competences for lifelong learning. European reference framework*. Luxembourg: Office for Official Publications of the European Communities.
European Union. 2009. *European year of creativity and innovation 2009*. http://www.create2009.europa.eu/index_en.html.
Feldman, D. H., M. Csikszentmihalyi and H. Gardner. 1994. *Changing the world: A framework for the study of creativity*. Westport, CT: Praeger Publishers.
Florida, R. 2003. *The rise of the creative class*. Melbourne: Pluto Press.
Hall, S. 2010. "Foreword". In *Cultural expression, creativity & innovation*, ed. H. Anheier and Y. R. Isar, ix–xi. London: Sage.
Kaufman J. C., and J. Baer, eds. 2006. *Creativity and reason in cognitive development*. Cambridge: Cambridge University Press.
Kuhn, T. S. 1962/1970. *The structure of scientific revolutions*. Chicago: University of Chicago Press.
Merton, R. K. 1961. Singletons and multiples in scientific discovery: A chapter in the sociology of science. *Proceedings of the American Philosophical Society* 105 (5): 470–86.
Morelock, M., and D. H. Feldman. 1999. "Prodigies". In *Encyclopedia of creativity*, Vol. 2, ed. M. Runco and S. Pritzker, 449–56. San Diego: Academic Press.
Nussbaum, B (2008) "Innovation" is dead. Herald the birth of "transformation" as the key concept for 2009. *Bloomberg Businessweek*. http://www.businessweek.com/innovate/NussbaumOnDesign/archives/2008/12/innovation_is_d.html?campaign_id=rss_blog_nussbaumondesign.
Plucker, J. A., ed. 2003. *Human intelligence: Historical influences, current controversies, teaching resources*. http://www.indiana.edu/~intell/simonton_interview.shtml.
Potts, J. 2007. What's new in the economics of arts and culture? *Dialogue* 26 (1): 8–14.

Pratt, A., and P. Jeffcut. 2009. "Creativity, innovation and the cultural economy: Snake oil for the twenty-first century?" In *Creativity, innovation and the cultural economy*, ed. A. Pratt and P. Jeffcut. New York: Routledge.

Redhead, S. 2004. Creative modernity: The new cultural state. *Media International Australia* 112: 9–27.

Sawyer, R. K. 2006. *Explaining creativity: The science of human innovation*. New York: Oxford University Press.

Simonton, D. K. 1994. *Greatness: Who makes history and why*. New York: Guilford Press.

———. 2003. Scientific creativity as constrained stochastic behavior: The integration of product, person, and process perspectives. *Psychological Bulletin* 129 (4): 475–94.

———. 2004. *Creativity in science: Chance, logic, genius, and zeitgeist*. Cambridge: Cambridge University Press.

Sternberg, R. J., E. L. Grigorenko and J. L. Singer, eds. 2004. *Creativity: From potential to realization*. Washington, DC: American Psychological Association.

Tusa, J. 2003. *On creativity*. London: Methuen.

UK Department for Culture, Media and Sports. 2008. Creative Britain: New talents for the new economy. http://www.culture.gov.uk/reference_library/publications/3572.aspx.

2   National Innovation Systems and Creativity

Jonathan West

A common theme of the chapters in this book is that creativity and innovation are not identical, and that innovation does not necessarily stem from creativity. That is, creativity alone does not necessarily produce innovation, and innovation does not necessarily depend on creativity. Innovation flows from the systems that direct and discipline creativity towards defined practical goals. Another theme in this book is that individuals do not innovate in isolation—they collaborate and draw upon externally generated resources. Nor does creativity itself take place in isolation.

The significance of these two insights is that to find the roots of innovation and creativity we must understand the context in which they occur and the influence of the social-systemic context on the direction and level of innovation. Other chapters discuss the impact of organisation and management on innovation and creativity. This chapter will discuss what research in the disciplines of economics and organisation theory has revealed about the role of national systems—sets of economic, political, social and cultural institutions—in determining the direction and pace of innovation and creativity. It then considers some implications for policy.

WHAT IS AN "INNOVATION SYSTEM" AND WHAT DOES IT DO?

Before discussing innovation systems in any detail, we should first define the term "innovation" as it pertains to the world of industry and commerce. An influential definition is "the processes by which firms master and get into practice product designs and manufacturing systems that are new to them" (Nelson 1993, 4). This definition can be expanded to include novel organisational practices and service delivery. Note that this definition also encompasses the commercialisation and introduction into practice of new ideas, not merely invention; that it includes the products and the processes by which they are produced; and that it does not limit innovation to the first introduction of a new idea.

A national innovation system can, in turn, be defined as the "set of institutions whose interactions determine the innovative performance, in the sense above, of national firms" (Nelson 1993, 5). These institutions go beyond factor and product markets—essentially the determinants of price—to include political and social institutions; labour training and employment norms; laws governing financial markets and taxation; education; patent law; publicly sponsored research; as well as culture, history and values (North 1990). In essence, a national innovation system performs three vital social functions: it mobilises and allocates resources, it determines the appropriation and allocation of returns, and it manages the risk needed to undertake technological advances.

The term "system" does not imply that the complex interactions among the aforementioned elements need be either planned or deliberately created. Indeed, in most successful countries, centralised planning has either been deliberately eschewed or abandoned. Nonetheless, all successful countries have acted to shape the factors that comprise an innovation system, and some such as Singapore or Taiwan have introduced far-reaching efforts to create favourable institutional climates (Goh 1995). In an economic sense, the innovation system may be considered to be the social institutions through which societies shape and expand their comparative advantage.

While the economic theory of comparative advantage offers valuable insight into how nations might raise their productivity in the short term— by trading with others and specialising in the fields in which they enjoy relative advantage—it offers little guidance as to how such advantage might be constructed in the first place. In the days of David Ricardo, the theory's originator, comparative advantage was regarded as largely determined by primary factor endowment in land and climate: a gift of God. Ricardo famously suggested the example of Portugal specialising in wine and England in textiles. In the centuries since Ricardo, it has become obvious that in the manufacturing and service industries that have driven economic growth, comparative advantage is mostly a human creation, more a result of effort, skill, and organisational and institutional effectiveness than natural endowment.

HOW DO INNOVATION SYSTEMS DEVELOP AND EVOLVE?

The study of national innovation systems is thus the investigation of how comparative advantage is created over time. Unfortunately, economic theory has been of limited assistance in this research. Economics focuses on markets: interactions and transactions among individuals and organisations. Why some individuals, organisations, regions or nations are better than others at performing the tasks that matter in market transactions has been seen by most economists as outside the scope of economic theory, more a matter for historians, business analysts or organisation theorists.

Yet, at its heart, comparative advantage is about just such economic capability, the ability to meet human wants better than competitors.

Nor has public discussion assisted much in understanding the sources of comparative advantage. Media coverage of economic issues rarely focuses on capability. It tends to dwell on headline stories of managerial blunders and power struggles, mergers, acquisitions, business cycles, currency exchange and interest rates, taxes and fluctuations in energy prices. But none of these issues are fundamental. They are ultimately surface phenomena that contribute only to shallow capability: short-term pricing and cost issues. Separate from the dramas that surround these issues are the enduring factors that determine sustainable prosperity.

Because the media focuses on the shallow factors, so too often have political leaders. Politicians frequently ignore the deep capabilities that develop gradually over time and endure longer. These capabilities include accumulations of strategic resources and proprietary knowledge, which require organisational routines, employee commitment and superior problem-solving to be realised. Deep capabilities are those aspects of the economy that are difficult for others to copy and that support ongoing gains in competitiveness. To develop new capabilities—comparative advantage— we need to move from a static to a dynamic perspective.

The shift to a dynamic view of economic development is the focus of research into innovation systems. The study of national innovation systems begins with the recognition that innovation systems differ across nations, and that such variation is shaped by those nations' history and context. But study of national innovation systems is built also on an understanding of how advances in technological and economic practice actually happen, the key actors and processes involved and the demands on these actors.

Many important technological advances are associated with advances in science. A grasp of how science operates in different countries is therefore essential for understanding innovation systems, although innovation depends also on many other inputs and factors. In addition, innovation takes place mostly through organisations, usually firms, which organise their research and development activities and innovation projects, fund them and decide which to pursue to implementation and commercialisation. An understanding of characteristic forms of corporate and not-for-profit organisation, and the norms governing interaction within and among firms and their personnel (that is financial, product and labour markets), is therefore also essential. And, finally, companies and the institutions of science and research (universities, public research institutes and funding bodies) are structured and regulated by government. Therefore, understanding the role of government in fostering the national innovation system is necessary.

The processes participants use to innovate are governed in the first instance by the demands of the innovation process itself. Technological advance (as distinct from commercial innovation) sometimes proceeds through dedicated R&D laboratories and facilities, staffed by university-trained scientists and engineers and funded by firms, universities and government agencies. The creation and maintenance of research facilities and recruitment and retention of their staff is thus an important feature of any national innovation system.

The "high-tech" sectors that focus on R&D, however, generally account for only a small proportion of either innovation or growth. In most OECD economies high-tech manufacturing makes up less than 3 per cent of GDP. All OECD economies consist of a combination of large medium-technology and low-technology manufacturing industries (such as food and beverages, or fabricated metal products) and large-scale service activities (of which the largest are education, health and social services). Research reveals that so-called low- and medium-technology industries include significant proportions of innovating firms that develop new products and generate significant sales from new and technologically changed products (Smith and O'Brien 2008). Indeed, the very distinction between "high-" and "low-" technology industries is misleading, depending on an inadequate measure: the proportion of the industry's R&D performed internally by individual firms. A more useful classification contrasts industries with internal R&D, in which innovation is driven by R&D performed by companies, from distributed-knowledge industries, in which innovation is developed externally and then optimised internally. Some industries can be "knowledge-intensive" without a great deal of R&D performed by firms within the industry. An example of such an industry is agriculture, which in the developed nations has sustained innovation-driven productivity advance in excess of most other industries, but in which the overwhelming majority of R&D is performed by public-sector organisations and subsequently diffused to the private sector (farms).

Innovation follows different paths in different industry sectors (see also Marceau, Chapter 3, this volume) with divergent approaches, methods and results. Key dimensions of difference across industry sectors include the relative propensity to new company formation; product versus process focus; and internal versus external knowledge sourcing. In some industry sectors innovation primarily takes the form of new company formation (for example, software and certain fields of electronics); in others, it manifests through the activities of already-existing large companies. In some sectors innovation is primarily product-focused; in other sectors (for example, metals and energy production) it is process-focused. In yet other sectors it is science-research-based (for example, pharmaceuticals); in others it is marketing-focused (many consumer goods).

SCIENTIFIC CREATIVITY AND INNOVATION

The relationship between creativity in science and innovation is thus complex, and varies from scientific field to field, and nation to nation. The direction of causality is not unidirectional. In some instances, new science gives birth to new technology and commercial innovation. This is the simplest picture, and the one that advocates of more spending on science and education usually have in mind. Here, innovation is seen as the commercialisation of inventions made in scientific labs, and it follows in these sectors that an important emphasis of policy ought to be to encourage researchers to pursue industry-useful research and to develop mechanisms to take inventions through to commercialisation.

Just as often, however, commercial innovation gives impetus to new science, or draws upon existing science in ways its originators could not foresee. Serendipity plays a major role in the relationship between science and innovation. Innovations often spring from applications of science that are quite unexpected by their scientific discoverers. In addition, new technologies often precipitate new science aimed at a deeper understanding of what seems to work, and improving it. Often, a commercial production process is not simply a scaled-up version of lab procedures, but an entirely new process, the result of considerable scientific and engineering work. For example, a modern drug-production process, especially in biotechnology, is not a scaled-up version of the vacuum flasks and reactors in which discoveries were originally made. A new process for drug production must often be invented.

An innovation system must therefore be understood as creating supply of science as well as supporting demand for science. The vehicles for science-based innovation are existing companies looking for new products and solutions as well as new companies created to commercialise discoveries. The balance between existing and new companies in this equation varies by industry, field of science and country.

But in every field of science, innovation depends on experimentation (West and Iansiti 2003). Once problems have been identified and defined, whether on the supply side by engineers or scientists or on the demand side by marketers, a set of potential solution options must be offered, and a strategy to test the options must be developed so as to eliminate those that are too costly or have an unacceptably low probability of success. Seldom can commercial innovators be assured in advance that all the pieces will fall into place for the new product or process: that the projected technology will work as expected, that a market will be found and that managerial and technical staff will be capable of meeting the challenges and obstacles encountered. Innovation is always, therefore, both "inefficient" in that activities must be undertaken that will probably fail—and risky—in that it will possibly yield little or no value.

Indeed, risk is the defining challenge of innovation. And the locus and intensity of risk varies by industry and technology (West 2004). In some sectors, the technology is likely to function as anticipated, but finding a sufficiently large market for it will be difficult. This is true, for example, in much information technology. In other sectors, a market is known to exist, but getting the technology to operate effectively and to scale will be problematic. This is often the case in drug development.

Sustaining science and the efforts of organisations to commercialise new technology thus requires investment of considerable resources, at substantial risk, often over long periods. An effective national innovation system will therefore include a means to mobilise these resources, to allocate them to risky undertakings, to evaluate the progress of innovation projects and to eliminate those with unacceptably low prospects of success.

Diversification is often critical to risk management. Once all efforts have been made to reduce risk by careful consideration of options and selection of capable research management teams, the classic way to manage risk is to diversify it: to pursue multiple projects in the hope that successes will more than offset the inevitable failures. Sometimes such a diversified portfolio is managed within the company as a portfolio of projects. At other times risk is managed through a portfolio of new companies, as in the case of venture capital firms. Ultimately, both risk management strategies rely on diversification.

The outcome of these strategies and investments is a system whose elements must be coordinated to enable innovation and creativity to function effectively. Organisations (including firms, non-profit institutions and government-sponsored agencies) must align their activities with laws and norms regulating how organisations operate and interact and with the process of technological innovation itself.

It is important to note that while the challenge originates in the nature of the innovation process itself, the solution is often addressed in different ways. The exact combination of institutions, strategies and practices that characterises any successful national system will depend on the specific history, culture and values of that nation.

FINDINGS FROM RESEARCH IN INNOVATION SYSTEMS

The study of innovation reveals many ways to fail but perhaps more surprisingly also more than one way to succeed. Table 2.1 illustrates in schematic form characteristic dimensions of the innovation systems in three successful innovating nations, the United States, Japan and Singapore. Based on my own analysis, along with that of others, I identify six important dimensions (tasks which national innovation systems must perform), as described earlier: investment mobilisation, capital allocation and risk management, basic research location, commercialisation path, professional labour market and primary value capture mechanism. Comparison among the countries reveals considerable variance in how these tasks are performed, here in all instances successfully.

Several important implications emerge from this work. The first is that innovation is "systemic" in the full sense. Many researchers observe that

Table 2.1   Three National Innovation Systems

	US	Japan	Singapore
Investment mobilisation	Low domestic savings; capital import	High domestic savings	Very high savings; government forced
Capital allocation and risk management	Capital markets; venture capital	Corporate retained earnings; banks	Government; government-linked corporations
Basic research location	Universities; government-sponsored labs	Large corporations	Government-sponsored labs
Commercialisation path	Start-up venture capital—Initial Public Offering	Large corporation or spin-off within keiretsu	Sell to foreign-owned multinational corporation; government-linked companies
Professional labour market	Broad and deep	Narrow and shallow; lifetime employment	Developing
Primary value capture mechanism	Equity; intellectual property	Production; corporate earnings	Wages; some taxes

national innovation systems are not simply pigeonhole lists of institutional structures and "good" policies, in which possession of more "best practice" features means more innovation. Rather, they are coherent systems in the full meaning of that word, in which interaction among elements within the system determines performance. With four out of six features of an effective system, for example, a nation does not gain 66 per cent of the benefit but often none. Moreover, it is often not feasible to mix and match features from different national systems, combining the best from A with the best from B. What works well when country A adopts a combination of features might be ineffective in other countries.

The second implication of this research is that basic knowledge creation often occurs outside the market. This is one of the best-established results in the field (see also Joshua Gans, this volume). As long ago as 1962, Kenneth Arrow showed that a "competitive system" (by which Arrow meant a freely functioning market) will fail to achieve "an optimal resource allocation in the case of invention" (1962, 615). He argued that a free market, left to its own devices, will allocate fewer resources for invention (which he defined as the production of knowledge, not the commercialisation of invention) than would be desirable. The essential reason is that individual participants in a fully competitive market cannot capture returns sufficient to justify bearing the risk. Arrow concluded that:

For an optimal allocation to invention it would be necessary for the government or some other agency not governed by profit-and-loss criteria to finance research and invention. In fact, of course, this has always happened to a certain extent. The bulk of basic research has been carried on outside the industrial system, in universities, in the government and by private individuals. (624)

Indeed, all successful innovating nations have found a mechanism to supplement under-investment by private firms in research and invention. Many, of course, provide generous funding to universities; Japan and other East Asian countries have created mechanisms such as the keiretsu and lifetime employment that allow firms to capture the benefits of riskier basic research. Evidence from a large-sample statistical study supports Arrow's hypothesis (Furman, Porter and Stern 2001). The study examined the innovation output of seventeen industrialised countries, and related the output to a variety of resource and contextual factors. The results were clear: government resource commitment, especially to education and research, as well as government policy, mattered a great deal:

We find that while a great deal of variation across countries is due to differences in the level of inputs devoted to innovation (R&D manpower and spending), an extremely important role is played by factors associated with differences in R&D productivity (policy choices such as the extent of IP protection and openness to international trade, the share of research performed by the academic sector and funded by the private sector, the degree of technological specialisation, and each individual country's knowledge "stock"). (Furman, Porter and Stern 2001, 924)

The study noted that between two-thirds and 90 per cent of the overall variation in national innovation (measured by patent output) could be explained by measures of R&D expenditure and total economy size, and found that a one percentage point increase in the share of resources going to higher education was associated with an 11 per cent increase in the national innovation output. Significantly, the study found that "countries with a higher share of their R&D performance in the educational sector (as opposed to the private sector or in intramural government programs) have been able to achieve significantly higher patenting productivity" (Furman, Porter and Stern 2001, 925). This was especially true of countries that had increased their performance the most.

The third implication of innovation research is that innovation is risky and expensive and demands sustained commitment. Many, if not most, governments will seek to minimise that risk and commitment in favour of politically easier paths. For many nations, economic growth can be achieved by expanding existing industries and *not* attempting to innovate in new industries and programs. Certainly, while innovation can drive economic growth, it is not identical to it. A more "efficient" way to increase economic growth for many nations will be to apply well-understood technologies to existing industries. The problem with this approach is that as the nation reaches the technological frontier, growth declines.

But to enter industry sectors at the leading edge of technology—often characterised by high growth and high value added—requires direction of investment into areas of considerable risk, at least at the start-up of the industry. Taiwan's establishment of a semiconductor industry is a good example. Taiwan's semiconductor industry began in 1977, rather late in the history of the industry. The Taiwanese government's decision to make a belated entry into the semiconductor industry posed great risks for the nation. In the late 1970s, the industry was already dominated by powerful global companies based in the United States and Japan. Most industrial research was concentrated in the United States and Japan, as were education in the technology and markets for the products. Many thought the prospects for successful entry by Taiwan, a distant and poorer new player, did not look good.

By 2000, however, Taiwan's industry had emerged as the world's third largest in production behind the United States and Japan, and was rapidly closing the gap with these leaders. The industry had driven increases in Taiwan's productivity and living standards for two decades, growing at a cumulative rate of more than 10 per cent per year. How was this dramatic success achieved? One important factor was resource mobilisation. Taiwan's savings rate averaged about 30 per cent of GNP between 1969 and 1997, and household saving over the same period averaged more than 20 per cent (net household saving in many developed countries is around 2 per cent, and recently even lower). To gain high savings rates, the Taiwanese government pushed down private consumption, which decreased as a share of GDP from 74 per cent in 1952 to 47 per cent in 1987 (Scott 2000).

Taiwanese firms were supported to enter the semiconductor industry by government—at least until they could stand on their own feet. The Taiwanese government established Hsinchu Science–based Industry Park as a hub for the semiconductor industry and encouraged firms to move there. Although the small firms were privately owned, they received many inducements: attractive terms for establishing a business, taxation allowances, low-interest loans, matching R&D funds, special exemptions from tariffs and commodity and business taxes. All this required substantial government investment and ate up the country's savings. But Taiwan went further. The government established the Industrial Research Institute, with a 1996 budget of US$1 billion and six thousand employees, 75 per cent of whom were researchers and five hundred of whom held doctorates. The institute was charged with importing and developing semiconductor-relevant technology and then licensing it to private firms.

The Taiwanese government also provided venture capital for the first semiconductor firms, United Microelectronics Corporation and Taiwan Semiconductor Manufacturing Corporation (TSMC), and entered a joint venture arrangement to ensure TSMC was sustained (Mathews, Cho and Cho 2000). Only after fifteen years of government investment and absorption of risk did the first substantial private capital enter the industry. By 1995, Taiwan possessed twelve semiconductor fabrication facilities, with sales of about US$3.3 billion. By 2000, that number had jumped to US$17 billion, or approximately 5 per cent of Taiwanese GNP.

THREE RISK MANAGEMENT APPROACHES

The Taiwan example illustrates the scale of government resource mobilisation and commitment sometimes required to enter an entirely new innovation-based industry. In these circumstances, considerable risk must be assumed and managed. Every successful national innovation system has developed an effective risk management approach and all involve a mechanism for diversification. Three different approaches have proven successful in different contexts.

Most new businesses create a "me-too" product or service, incurring little risk (Bhide 2000). These businesses start small and remain small, although they can provide a prosperous return to an individual entrepreneur. Being relatively low risk, but with modest growth prospects, most such ventures are funded from personal resources or from family and friends. But while small "me-too" firms are numerous, they often have a relatively short lifespan. The typical entrepreneurial firm that grows into a large-scale and sustainably successful firm is somewhat more risky, though not initially larger scale. Most such firms take several years to define a niche in which they can gain a distinctive competence, and during that period their customers implicitly agree to share the risks involved. As Bhide shows, success for these firms is often based on out-hustling others with similar ideas, though obviously their rapid development is based on distinctive ideas. Such firms usually struggle for five or more years before they gain the competence that attracts formal venture funding. They are often financed also by a combination of personal assets and friends-and-family assets.

For larger and riskier undertakings, sources of capital that appear small in the overall picture assume much greater importance. Such ventures require substantial funding, beyond the resources of most individuals and especially of the inventors, are very risky and frequently require a longer period before novel ideas become a commercial product. To cope with the challenge, entrepreneurs must turn to investors who can diversify risk.

The three main vehicles for such investment are venture capital and private investors; large corporations, including banks; and government.

Different national innovation systems tend to favour one or other of these vehicles for entrepreneurial risk diversification. While all three approaches are employed in most countries, the particular mix and emphasis for obtaining investment is one of the defining features of national innovation systems. To summarise a large body of research literature: US and "Anglo-Saxon capitalism" typically relies more on venture capital; European "welfare-capitalism" relies more on government and banks; and Japanese "keiretsu-capitalism" relies more on large corporations (Berger and Dore 1996; Dore 2000).

The fourth finding from research in innovation systems is that the appropriate investment structure for innovation must match its risk structure, that is, larger and riskier projects need to be financed by larger and more diversified investing entities if risk is to be managed effectively. All innovation projects contain three basic types of risk:

- *Technical risk*, that is, whether the product, process or service will actually perform the intended function.
- *Market risk*, that is, whether a sufficiently large market can be found for the product.
- *Managerial risk*, whether the organisation attempting to innovate can assemble the leadership team required to bring the innovation to fruition.

Most venture capitalists attempt to remove, or substantially reduce, technological risk before they consider investment. Discussions between technological entrepreneurs and venture capitalists usually begin with "proof of concept": evidence that the device, software program or service actually works.

Venture capitalists focus mainly on managing market and managerial risk. In the fields in which venture capital has flourished—information technology, software and telecommunications—it is usually possible to demonstrate at the outset that the proposed concept is technically feasible and practical, at least in principle. The underlying physics and engineering are usually well characterised.

In life science, by contrast, once technical feasibility is established—for example, the research institute demonstrates that the drug is safe and effective against a cancer, with acceptable side effects—commercial success is virtually assured. Most life science projects and life science start-ups come into being precisely to determine whether the concept will work technically. In life sciences, with success rates as low as one in twenty, and minimum investments becoming very large (to take a potential drug through all phases of development and registration, for example, now costs more than US$700 million), the required size and diversification of the portfolio is beyond all but the largest firms. A portfolio of only twenty projects at US$700 million each would require a commitment of US$14 billion. For the riskiest life science enterprises— technologies based on new genetic discoveries—investment is usually beyond the reach of all but the very largest firms or government. It is not surprising, then, that even in the United States, only approximately five venture capital funds specialise in biotechnology, and the proportion of venture capital investment in biotechnology has actually declined in recent years. In life science, potential investors confront irreducible risk of all three kinds.

Fifth, innovation research has shown that value from innovation is increasingly captured by equity owners rather than wage-earners. In the past, nations might capture substantial value from innovations in the form of wages and taxes by ensuring that production activities employing these innovations took place within their borders. In the knowledge-based industries that often drive contemporary industrial innovation, however, little value might be captured as wages. This is because replication of a product design—that is, manufacturing and service delivery—is increasingly trivial in key industries. Value is increasingly obtained in the original design. Once one copy is perfected, making a million more poses little challenge and merely captures value for the design owner.

An example is computer software. Developing the first version of a computer program is highly skilled work, time consuming and usually well paid. Once the "design"—that is, the code—is perfected, replicating it is trivial. Distributing it over the Internet utilises little labour. And more and more industries are looking like software: the value is concentrated in intellectual property. Table 2.2 shows the proportion of added value captured by wages in several representative industries.

Table 2.2   Value Added by Function (%)

	Wages	Net operating surplus
Precision engineering	67	6
Specialty chemicals	67	9
Disk drives	24	58
Computers	11	89
Life sciences	7	91

Source: Author's estimate.

NATIONAL INNOVATION SYSTEMS IN OPERATION: THE SEMICONDUCTOR INDUSTRY

My own work on the dynamics of innovation in the global semiconductor industry illustrates many of the points made earlier about the integrated nature of innovation systems. I found that typical business organisations in the semiconductor industry in Japan and the United States had evolved along different paths, with each element of technology, institutions and organisational form supporting each other to form quite different systems (West 2000, 2001, 2002, 2004). Similar results have been found by other researchers in a range of industries (Dore 1994; Clark and Fujimoto 1991; and several of the studies in Imai and Komiya 1994). The differences observed in the United States and Japanese semiconductor industries are summarised in Table 2.3.

The two key factors in the semiconductor industry that underpinned these national differences in practice were the structure of markets for university-trained labour and the operation of national research systems.

As many researchers have noted, Japanese enterprises frequently extend "permanent employment" rights beyond management and shareholders to technical and shop-floor workers. As a consequence, the market for professional labour in Japan remains thin and shallow. Japanese firms typically induct university-trained personnel only upon graduation from university or college (Westney and Sakakibara 1986), and for all practical purposes do not recruit personnel later in their careers. Only in unusual circumstances would university-trained personnel in late career seek to join another firm, nor typically would any of the Japanese firms I interviewed seek to hire such personnel. By contrast, a vigorous labour market exists for university-trained personnel in the United States. US firms can and do recruit already trained and experienced personnel, and these professionals expect that career mobility and transfer to another company is an essential part of their development.

Table 2.3 Semiconductor Industry R&D Organisational Practice: United States and Japan

	Japan	*US*
Skills Acquisition and Retention	Experience-Based	Education-Based
Program scope and leadership	Loose	Tight
Program guidelines and timing	Implicit	Explicit
Task partitioning	Distributed	Focused
Resource allocation	Decentralised	Centralised
Experimentation capability and practice	Low-Medium	High

There are also important national differences in education, training and skill-formation systems. The US higher education system produces considerably more PhD-level graduates than does the Japanese system (Lynn, Piehler and Zahray 1988). Japanese firms expect to train employees themselves or to sponsor external training. US firms assume that skills acquisition is mostly an individual responsibility. Japanese firms invest in the skills of employees because they are confident of retaining those skills over time (Lynn, Piehler and Kieler 1993). Conversely, the relative underdevelopment of the graduate-level higher education system obliges them to do this.

These national differences in employment and training may help explain the observed differences in innovation project–organisation practice. Why, for example, do Japanese organisations distribute personnel and R&D budget resources more evenly among multiple sub-units, and not establish formal project-specific teams, with staff dedicated to a single project, as is more the case in the United States? The answer appears to lie in guaranteed employment continuity, which allows Japanese firms to pursue an experience-based approach to knowledge creation and problem-solving. Employment continuity also facilitates deeper organisational socialisation in Japanese firms, enhancing communication and coordination across sub-units. Japanese employees report that they develop strong relations with other employees of the same firm over many years, and are deeply familiar with each other's work style and preferences. This understanding improves communication and reduces problems of interfunctional and interdiscipline transfer of knowledge. Paradoxically, therefore, it may be that Japan's stronger organisation-specific integration and socialisation make teamwork easier but formal team structures less necessary. The result is that Japanese organisations experience less need for project-team-type organisation.

US managers work within a different set of organisational imperatives and opportunities. Confronted by the need to improve their innovation capability in the mid-1980s, many US firms decided to move away from their former functional and discipline-divided organisational structures. This structure had produced cost overruns, time delays and some poor-quality products. But the US firms were less able to build experience-based strategies. In the United States, with fluid markets for highly skilled labour and strong professional/discipline (rather than company) loyalty, most firms could not assume they would retain their experienced staff over the long term. Even the strongest firms, such as Intel and IBM, risked loss of key personnel in the event of business down-turn.

The US business context creates an environment favouring professional over organisational socialisation. US employees in the semiconductor industry were often more integrated into their professions as electrical engineers, solid-state physicists or semiconductor specialists than committed to their current company. Shallow organisational integration produces communication barriers and difficulties. US organisations commonly reported problems in building effective communication across internal organisational boundaries. This helps explain the problems widely believed to have pre-cipitated the delays and cost overruns that damaged the US firms' competi-tive position in the mid-1980s.

The shift to dedicated project-based teams in the 1990s, described in my studies, helped the US firms solve these problems. The shift was encour-aged by the ease with which US firms could recruit already highly trained personnel. The successful firms could hire well-educated scientists and engineers, either from the strong US graduate-school master's and PhD pro-grams or from other firms. These personnel were well trained in designing and executing experiments and possessed strong knowledge of scientific fundamentals in relevant fields, but often lacked on-the-job experience. These teams became increasingly central to the effort of product and pro-cess development. Thus, key elements of the US organisational mode—its focus on experimentation, tight project teams and centralised resource allocation and task partitioning—were all ultimately encouraged by the labour market constraints and opportunities within which US managers made their choices.

In sum, institutional context, incentives and constraints can explain the observed differences between Japanese and US organisational practice, especially those related to skills acquisition, project-team organisation, task partitioning, resource distribution and capacity building. Each com-ponent of the system fits the others. In neither the United States nor Japan would it be feasible to simply graft on elements of the other's system.

IMPLICATIONS

Discussion of the origin, functions and differences among national inno-vation systems suggests important messages for organisational practice in innovation and the role of governments in shaping national comparative advantage.

The following discussion focuses on policy implications. The single most important outcome from the research reviewed here is that government can exercise a major influence over the evolution of a nation's innovation system. But considerable care must be exercised to ensure that influence strengthens, rather than undermines, that system.

First, government should complement, not attempt to substitute for or compete with the activities of private companies. Government should aim to strengthen the market position and capabilities of private firms. The first principle is that government policy should "do no harm": it should not distort market incentives by attempting to substitute public-sector activity for those of the private sector. It should limit its domain to fields in which infrastructure, capability and resources are both demonstrably required to support innovation and in which it is not feasible for private companies or markets to undertake the specified activity.

Second, government policy should aim to exercise influence on industry sectors of sufficient weight and potential to matter. Innovation policy should address the needs of relatively large sectors. This implies that an innovation policy cannot be limited to "research-intensive" industries such as information technology, biotechnology or nanotechnology. It is well to recall that so-called high-tech industries (usually defined as industries with R&D/sales ratios of more than 4 per cent) make up only a small component of manufacturing in any economy, and an even smaller component of GDP.

In a modern economy, especially a small one, many of the goods and services citizens want can be obtained only from abroad; generating the income to pay for these imports depends on what the nation can sell to the world. The economic fate of nations can thus rest on a surprisingly narrow base of capability in a very few fields. Ensuring the long-term strength of these sectors ought to be a high priority for any community and its government.

Third, government should address the needs of these high-leverage industrial sectors based on an analytical understanding of their actual innovation systems and processes. As noted earlier, innovation follows different paths in different sectors. The distinctive innovation patterns across sectors must be understood if effective policy is to be developed and applied successfully.

These three policy implications suggest a final conclusion: the construction of comparative advantage through a national innovation system implies geographic and sectoral *decentralisation*. To promote capability and comparative advantage effectively, government policy ought to focus on sectors in which the economy specialises and be specific about where to locate and support facilities, institutions and resources to support these key sectors. Perhaps the final message from this discussion of research findings about national innovation systems is there can be no effective one-size-fits-all, "best" economic policy for the design and development of a national innovation system.

BIBLIOGRAPHY

Arrow, Kenneth. 1962. "Economic welfare and the allocation of resources for invention". In *The rate and direction of inventive activity*, ed. Universities-National Bureau, 609–26. New York: Princeton University Press.
Berger, Suzanne, and Ronald Dore, eds. 1996. *National diversity and global capitalism*. Ithaca, NY: Cornell University Press.
Bhide, Amar V. 2000. *The origin and evolution of new businesses*. New York: Oxford University Press.
Christensen, Clayton M. 1997. *The innovator's dilemma: When new technologies cause great firms to fail*. Cambridge, MA: Harvard Business School Press.
Clark, Kim B., and Takahiro Fujimoto. 1991. *Product development performance: Strategy, organisation, and management in the world auto industry*. Boston: Harvard Business School Press.

Dore, Ronald. 1994. "Japanese capitalism, Anglo-Saxon capitalism: How will the Darwinian contest turn out?" In *Japanese multinationals: Strategies and management in the global kaisha*, ed. Nigel Campbell and Nigel Burton, 9–30. London: Routledge.

———. 2000. *Stock market capitalism: Welfare capitalism: Japan and Germany versus the Anglo-Saxons*. New York: Oxford University Press.

Furman, Jeffrey L., Michael E. Porter and Scott Stern. 2001. The determinants of national innovative capacity. *Research Policy* 31 (6): 899–933.

Goh, Keng Swee. 1995. *Wealth of East Asian nations*. Singapore: Federal Press.

Imai, K., and R. Komiya, eds. 1994. *Business enterprise in Japan: Views of leading Japanese economists*. Cambridge, MA: MIT Press.

Lazonick, William, and Jonathan West. 1998. "Organizational integration and competitive advantage: Explaining strategy and performance in American industry". In *Technology, organization, and competitiveness: Perspectives on industrial and corporate change*, ed. Giovanni Dosi, David J. Teece and Josef Chytry, 247–80. Oxford: Oxford University Press.

Lynn, Leonard, Henry Piehler and M. Kieler. 1993. Engineering careers, job rotation, and gatekeepers in Japan and the United States. *Journal of Engineering and Technology Management* 10: 53–72.

Lynn, Leonard, Henry Piehler and Walter Zahray. 1988. *Engineering graduates in the United States and Japan: A comparison of their numbers, and an empirical study of their careers and methods of information transfer. Final Report to the National Science Foundation, grant SRS-84099836*. Washington, DC: NSF.

Mathews, John A., Dong-Sung Cho and Tong-Song Cho. 2000. *Tiger technology: The creation of a semiconductor industry in East Asia*. Cambridge: Cambridge University Press.

Nelson, Richard, ed. 1993. *National innovation systems: A comparative analysis*. New York: Oxford University Press.

North, Douglas. 1990. *Institutions, institutional change and economic performance*. Cambridge: Cambridge University Press.

Smith, Keith, and Kieran O'Brien. 2008. *Innovation in Tasmania: An innovation census in an Australian state. Australian Innovation Research Centre Working Paper*. Hobart: Australian Innovation Research Centre.

Scott, Bruce R. 2000. *Taiwan: Only the paranoid survive. Harvard Business School case study 5–700–039*. Boston: Harvard Business School.

West, Jonathan. 2000. Institutions, information processing, and organization structure in research and development: Evidence from the semiconductor industry. *Research Policy* 29 (3): 349–73.

———. 2001. "Two best ways? Institutions and the evolution of the US and Japanese semiconductor industries". In *Comparative studies of technological evolution: Research on technological innovation, management and policy*, ed. Robert A. Burgelman and Henry Chesbrough, 37–59. New York: JAI Press.

———. 2002. Limits to globalization: Organizational homogeneity and diversity in the semiconductor industry. *Industrial and Corporate Change* 11 (1): 159–88.

———. 2004. Financing innovation: Markets and the structure of risk in non-replication economics. *Growth* 53: 27–43.

West, Jonathan, and Marco Iansiti. 2003. Experience, experimentation, and the accumulation of knowledge: An empirical study of the evolution of R&D in the semiconductor industry. *Research Policy* 32: 809–25.

Westney, D. Eleanor, and Kiyonori Sakakibara. 1986. "The role of Japan-based R&D in global technology strategy". In *Technology in the modern corporation*, ed. Mel Horwitch, 315–30. London: Pergamon.

3 Innovation and Creativity in Industry and the Service Sectors

Jane Marceau

The study of innovation in industry draws on insights from several social science disciplines: economics, sociology, economic geography, political science and management. The economics of innovation developed as a sub-discipline based on the work of pioneers such as Schumpeter (1942/1975), who coined the term "creative destruction" to describe the transformation that accompanies radical innovation under capitalism. He was followed by the economists Arrow (1962), Nelson and Winter (1982), Freeman (1982), Lundvall (1992) and Metcalfe (1998).[1] Sociology provides insights on power relationships between firms and between governments and firms (e.g. Orru, Woolsey-Biggart and Hamilton 1992; Smelser and Swedborg 1994). Economic geography uses location analysis to map innovation in firms linked in networks and clusters (Freeman 1991) and the relationships between participants in local, national and global supply chains (Sloan et al. 2005). Economic geography also examines the significance of "spaces" for innovation (e.g. Piore and Sabel 1984; Saxenian 1994). Political science offers insights into public policy for industry development and the impact of regulation (Stewart 1994). The discipline of management contributes ideas about how to organise for "innovation" at the level of the firm and public research organisation (e.g. Granovetter 1973; Lam 2005; West and Dodgson, Chapter 12 this volume). My discussion of innovation in different industry and service sectors will draw on insights from several of these social science disciplines.

The first section of the chapter describes some approaches to defining types of innovation, levels of innovation at different "levels" and sectors of an economy and the background conditions which underpin different types and levels of innovation. The second section of the chapter describes and illustrates a "product system" approach for examining the determinants of innovation in a particular industry or sector of an economy. The aim is to illuminate the information exchange and power relationships between the main players in a particular industry as well as the drivers of transformation in the sector, e.g. new technologies, so as to assist policymakers as they attempt to improve innovation.

DEFINITIONS AND APPROACHES

"Innovation" in the productive sector, whether in manufacturing or services, is usually defined as the creation of novelty of economic value. This translates into viewing innovation as the creation of new products and services, as change in the processes of producing these products and services and as organisational change, including new work practices. Innovation in industry may also mean using new methods to produce services and the creation of "product-service packages", either as "generics" for a market or to solve particular client needs (Marceau, Cook and Dalton 2002). An example of a new "product-service" package is the set of services provided as part of the "deal" or "bundle" when purchasing a mobile phone.

Both creativity and invention can be found in innovation. Here I define "creativity" as the creation of new ideas or a recombination of existing knowledge that has no immediate or particular market drivers. Invention is a new product (usually) or service whose economic value has not been tested in the market or whose value has been tested but has been found deficient and the new product remains unused. In this sense, creativity is similar to invention. Creative people are viewed as drawing on their thinking and research to find solutions to "problems", with an emphasis on thinking "outside the box", "playing" with existing products or services (cf. Dodgson, this volume) to find new ideas for product or services and new understandings of how "things work". Creative people are also those who can recognise new knowledge and apply it to address both scientific and product-related problems.

Creativity is difficult to define in relation to innovation. It is clear that successful innovation depends on creative people as new products and processes require people to think outside the box. This is especially true for radical innovation where new products, processes or services combine new and older knowledge in particularly novel ways. The creative spark is also involved in designing the organisational forms to improve production and sales. The creative leap involved in thinking of and about new knowledge combinations is seldom examined by innovation theorists. Accordingly, even in the management literature, creativity and innovation tend to occupy different conceptual spaces, as argued by Mark Dodgson in this volume.

Innovations may be "radical" or incremental. Some innovations change fundamentally the kinds of products produced, for example, computers and automobiles, while some are small improvements on existing products or manufacturing processes. Incremental innovations are far more common than radical innovations as most firms prefer to stay on familiar ground and make only small changes to minimise risk. Radical innovations, such as those in the information and communications technology, which begin life as products, may lead to major process changes; examples are computer-aided design and computers in manufacturing. Some new technologies begin as scientific breakthroughs—examples are biotechnology and nanotechnology—and take considerable time to become the platform for new products.

Modern economies tend to comprise a mix of firms making radical and incremental innovations. Firms may move between radical and incremental innovation as the platform technologies mature and the business environment changes. Radical innovation is more likely to stem from R&D, while incremental innovation, including process innovation, is more likely to stem from customer suggestions and feedback or by experience with the product. The different kinds of innovation all necessitate "creativity" as they involve doing things differently. The new business models essential for commercial breakthroughs also require creative thinking (an example is the Just-in-time operations management system), but often take time to emerge in the marketplace.

The so-called "high-tech" industries are considered significantly more innovation-intensive than their "low-tech" counterparts, but recent studies have shown that the so-called low-tech sectors often use leading-edge technology for product and process innovation (von Tunzelman and Acha 2005). "Old" industries, such as mining and agriculture, are in fact highly knowledge- and R&D-intensive, and have become constant innovators through the application of new science and new business models, including the outsourcing of key activities. The leading mining firms, for example, often outsource mine tunnelling operations to engineering and construction firms with "cutting-edge" expertise in tunnelling. This is an example of how links between different industries can transform operations in another.

There are many models of the innovation process. The linear model (knowledge–push) suggests that new knowledge moves directly from its creators in public-sector research organisations into commercial hands. Over time the linear model has been modified to include factors such as market pull, feedback loops, organisation of the firm, knowledge management and links with outside organisations. The new generation models point to distributed or "open" forms of innovation and establishment of networks to share knowledge and capabilities of several organisations (global and local), permanent and transient (Chesborough, Vanhaverbeke and West 2006; Bessant and Venables 2008). In the new generation innovation model, new IT-based "innovation technologies" enable new product and process simulation, rapid prototyping, team design and rapid steps to manufacturing through a process Dodgson, Gann and Salter (2005) refer to as "think, play, do" (see Dodgson, this volume).

Creativity plays its clearest role in the new generation models. The opportunity for highly creative ideas based on multiple sources of knowledge and perspectives to become an innovation grows as innovation becomes a more open, distributed and networked activity. As the new models of innovation take hold, creativity may attract greater attention among writers on innovation. With changes in the organisation of innovation we may see a deeper examination of creativity in firms and organisations.

COMPONENTS OF INNOVATION SYSTEMS

Innovation in an industry is dependent on the context (e.g. nation, region and city) in which industrial firms and related institutions function. Here I describe the components frequently identified as critical for innovative capability and performance and how they can be understood as elements of an integrated system.

Figure 3.1 depicts eight major "macro-level" components of a national system of innovation (cf. Freeman 1987). The eight components are like pieces in a jigsaw which fit together. The components are: development of human capital through education and training; suitable organisation of a nation's science (including social science) systems and of the intellectual property resulting from it; the rapid diffusion of new knowledge; investment in R&D to develop new technologies for production and sales, including platform technologies such as IT; availability of business services, especially to assist firms to access new knowledge and use new technologies; provision of venture capital to support risky but promising ideas for start-up companies and loans for company expansion; and trade opportunities.

The figure does not indicate which elements are the key drivers of change or innovation in a nation. This is because analysis shows that different nations have different configurations of each component and what drives innovation in one nation may be different in another (see also West, Chapter 2, this volume). The relationships between components in an innovation system are complex and dynamic. For example, an increase in government investment in R&D might produce a spectacular boost in innovation in

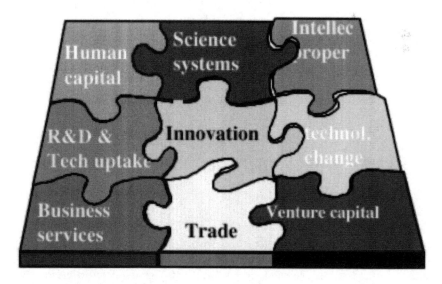

Figure 3.1   Eight components of an innovation system.

Taiwan but modest gains in North Korea. It depends on government regulation of IP, where the R&D is being directed and, of course, the expertise of the scientists doing the R&D.

Much of what is possible in a given national innovation system is affected by the industrial structure of that nation. By "industrial structure" I mean the characteristics of sectors in the economy, the technological level at which firms operate in key sectors, the size and international orientation of key players (e.g. large multinationals and small locals) and their relationships to each other through supply chains, knowledge generation and communication channels. Not all firms and ideas require the same components and policy frameworks. However, there is evidence from OECD studies and the work of Michael Porter (1990) and others that firms do best when they have large and demanding customers, have well-functioning supply chains, good access to R&D and scientific-technical skills and when they collaborate as well as compete.

The components in Figure 3.1 can be therefore understood as the "background conditions" of a national innovation system as well as the elements of the system. International comparisons show that some national innovation systems function better than others (OECD 2002, 2003). The link between innovation and sustained competitiveness in world markets, especially among manufacturers, has given policymakers a keen interest in national innovation policies, especially in regard to investment in sectors such as science and education. There is a debate over whether and how policymakers in any jurisdiction can affect levels of innovation. The literature suggests that policymakers should consider all eight components shown in the figure as critical for national innovation performance.

In brief, innovation performance at the level of industry sector or firm can be understood as the outcome of a system of relationships between organisations (firms) and elements of the environment in which they operate. These elements are linked together and together may impede or encourage innovation performance. Some are part of national innovation systems while others are more regional or local. The city as an innovation system is the subject of my other chapter (Marceau, Chapter 4, this volume).

INNOVATION IN INDUSTRY: LARGE FIRMS, SMALL FIRMS, HIGH TECH AND LOW TECH

What determines the level of innovation across different industries? International surveys (OECD 2002, 2003) have measured the levels of innovation activity undertaken by companies and suggest that expenditure on R&D is a critical differentiator for success. The surveys show significant differences in levels of innovation-related expenditure between nations, industry sectors (high versus low tech) and between firms of different sizes. High-technology firms, such as biotechnology, are innovation-intensive and spend up to 10 per cent or more of turnover per annum on R&D; science-based start-up firms may spend considerably more.

Observers do not agree, however, on whether size of an enterprise affects innovation level. Large firms have most R&D capacity but may be reluctant to take the risks associated with innovation, especially radical innovation. Many large firms prefer to rely on internal reorganisation to better integrate and use the knowledge they have already produced rather than invest heavily in their own R&D facilities and operations. Large firms also monitor the ideas and knowledge produced by smaller competitors and may then follow a strategy of merger and acquisition or collaboration. In this case, large firms are relying on others to creatively produce or integrate new knowledge. In more "technical" sectors, such as manufacturing, firms with large R&D departments have substantial technology and market-scanning ability and therefore greater knowledge-absorption capacity as well as the funds and staff resources to produce the new product efficiently and to manage the inevitable risks. While some writers claim that most R&D-related innovations come from large firms, others suggest that new, small, knowledge-based technology firms are the most innovative because they are more flexible. Where R&D is not the primary basis of innovation, as in smaller enterprises, flexibility and speed to market may be the driver of innovation.

Many firms obtain their innovation ideas externally. Studies have shown that customers are the single most important source of innovation knowledge, followed by suppliers and competitors (see e.g. von Hippel 1988; Marceau 1999). Maintaining close links with customers as end-users reduces innovation risk and smoothes the innovation process, especially in regard to product design. Some observers suggest that the most successful firms rely on multiple sources of innovation ideas and seldom a single source (Hyland, Marceau and Sloan 2006), especially, of course, if their markets are differentiated and consist of more than a few large clients.

The size of a firm, however, does not alone determine the degree and direction of innovation activity. The type of technology central to the firm, the stage of development of that technology and the type of market in which a company operates also shape decisions about innovation and investment in new products (Whitley 2000). Some recent studies attempt to identify the different sources and "packages" of influence (Hollenstein 2003). There is likely to be considerable industry variation as, for example, biotechnology firms may behave very differently from, say, metals manufacturers, at least in the early stages of development.

There are many elements important to success in innovation in manufacturing which is the subject of many of the studies referred to earlier. Recently, interest has turned to innovation in the service industries (e.g. hospitality, travel, tourism) and to innovation in industries such as building and construction, which integrate innovations from a range of other industries, e.g. design, service and manufacture.

The literature in the management and organisation field focuses on innovation at firm or inter-firm (supply chain or networks) level, what might be termed "micro-level systems of innovation". At this micro level of analysis, the focus is on internal aspects of organisational activity, leading to detailed work on enterprise processes of innovation, the management of innovation and management of knowledge, the management skills and technologies required for innovation and variations in innovative capacity of firms of different size and sector. West (Chapter 12, this volume) discusses some company strategies for innovation, especially in regard to information flow (see also Connell et al. 2001).

Not all analysts agree on the critical factors that determine the innovation success of different firms. Some focus on a firm's R&D and technological absorption capacity, while others suggest that market and customer orientation and links with other R&D-intensive organisations (both public and private) are key. The notion of "systems" of innovation at firm level might imply, perhaps inadvertently, that systems may be more important than the creativity of innovators for development of new products and processes. It is clear that "creativity" does not flourish in companies that impose tight control on employees rather than encourage cooperation and long-term loyalty to the firm (Collins and Porras 1994). And creativity needs support from across the firm. Critical support for creativity tends to reside in "leadership" by managers, especially at the senior level, who "champion" new ideas as they undergo rigorous scrutiny when firms consider provision of resources to take the ideas further. Managerial support is only part of the creativity process, however. Some firms take "creative people" out of the everyday business environment and provide time, incentives and organisational space for them to spark and develop new ideas. The chapter by Dodgson (this volume) shows several ways in which this is done in firms working in different industries.

MAPPING PLACES OF INNOVATION AND CREATIVITY: THE "PRODUCT SYSTEM" APPROACH TO INNOVATION IN INDUSTRY SECTORS

Over the past decade together with colleagues in the Australian Expert Group in Industry Studies (AEGIS),[2] I have developed a "product system" approach to examining innovation in a particular industry. The approach is concerned with identifying places in production systems where innovation is most likely to occur as well as the obstacles to more sustained or radical innovation and creativity. This approach to analysis of the innovation processes of different industries examines the inputs to innovation and maps the flows of knowledge as well as patterns of power and influence critical to innovation between participants in a product system. The flow of knowledge includes formal knowledge, such as new scientific results and technological novelty, as well as informal (tacit) knowledge gained from learning by doing.

The product system approach to innovation in industry "sectors" includes firms, R&D organisations, governments and customers (the latter are critical to innovation across many industries) as key participants ("players"). The approach draws on studies of the importance of user–producer relations in innovation (e.g. Lundvall 1992), the influence of regulation on new market development and inter-firm relations in supply chains as a stimulus and diffusion mechanism for innovation.

The product system approach as developed so far involves, first, identification of the key players in an industry innovation system and, second, investigation of the relationships between the key players on two dimensions or variables: (a) flow of information and knowledge between the players (but which, of course, can signal influence) and (b) the power or control relationship between the players, i.e. are there dominant players? Are there powerful coalitions? (A related dimension is collaboration–competition between players in a product system). Clearly, information flows and power-influence are related dimensions but are not identical. To analyse the two sets of relationships in a product system we use data based on interviews and statistical information gathered from the key players in several industry sectors (see Marceau 1999).

The first task to study a product system is to identify the players in the system. In our research the AEGIS group identified four sets of key players. They are: users (clients/customers), producers, R&D providers (and sometimes training institutions) and regulators (e.g. policymakers who determine the rules of the game and the accrediting agencies). The importance of these four main players has been identified in the literature on innovation (see e.g. Gallouj 2002 on service industries). It should be noted that of course there are other players who take key roles in particular industries, e.g. trade unions in the transport and waterfront industries. And as we shall see in case examples of industry sector innovation, there are other players in the product system who come into the analysis. The second step is to develop a generic model against which actual flows of information and patterns of power and influence may be subsequently observed and mapped. A third step is possible, and that is to compare the level of agreement or consistency between how the key players report the information flows and power collaboration relationships from their own perspective.We also make assumptions about what would constitute positive and constructive relationships on the two dimensions between the players in reference to the "open innovation" systems paradigm (Chesbrough, Vanhaverbeke and West 2006). First, we assume that plentiful and more or less equal flows of information exchange between the key players are conducive to greater innovation (model 1). Second, we also assume that in general, equivalence of power and influence (or use of it) between the key players is conducive to greater innovation (model 2), but note there are exceptions. We also assume

that conflict and tension will arise when there is imbalance between players in information exchange and/or power-influence relations. Conflict and
tension can be destructive to creativity and innovation but depending on
the reason and magnitude can also be a spur to change and to innovation.

The literature on continuous and open innovation indicates the benefit
of more or less parity of information and influence flows between the key
players for a product system. Gross imbalance (one-way flows of information and power) or, worse, no flow (a moribund system) is not conducive
to innovation.

Figure 3.2 is a simple "map" of the information exchange relationships
between four key players (customers/users, producers, regulators, etc.) in
a product system. This is a "perfect" map of information and knowledge
exchange. All four players are linked equally in information and knowledge exchange across all combinations. We would predict much innovation, unless of course the four players spend all their time informing and
not implementing! Few real-life situations approximate equality of information flows and influence. However, under some conditions, such as rapid
technological change or a dramatic shift in the regulatory environment (e.g.
removal of tariff protection), the industry may rely on powerful dominant
players to shift the organisational and operational arrangements and push
suppliers into compliance in order to survive.

Under other conditions, the major players may be the only ones able
to negotiate and coordinate the necessary but complex changes in relationships between all the players in the industry (e.g. with the regulators

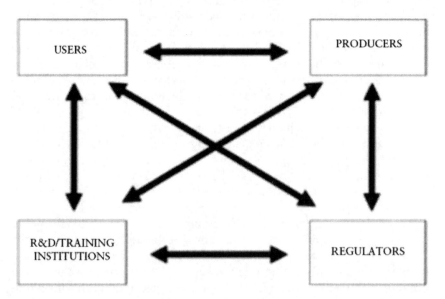

Figure 3.2   Model of equal information flows between four key players in an industry sector.

and the training institutions). Tensions may arise between the "creative" elements of the industry (who value their autonomy and independence) and their "innovative" counterparts who place a premium on business results and industry survival. In practice at any time some players dominate the activities of the industry. The case examples have been selected to focus on industries with significant creative elements—architecture and engineering in the building industry and fashion design in the clothing industry.

Our mapping of an industry product system has implications for policy. For example, the map might reveal perceptions and evidence of too little input from customers or from R&D organisations (information dimension) or too heavy-handed a policy and regulatory compliance framework from government (power relations). In each case this pinpoints the location of an obstacle to innovation. In interviews we look at information flow and degree of influence from the player's point of view. Thus, one player (producers) might claim there is little input from research technology organisations, and another reports there is too much interference by powerful regulators. The "maps" thus generated are slices of subjective reality designed to highlight issues and dynamics which industry and policymakers may wish to address for improving innovation.

It is important to emphasise that our maps of information flows and power relations are representations based on evidence we obtain from key players in an industry sector. The maps must be seen as an aid to examine the dynamics in relationships between players in an industry. They are simply a graphic representation based on research evidence to develop insights and formulate propositions for policy.

The following section describes several case examples taken from detailed case studies prepared by the AEGIS group (Marceau 1999) to illustrate the dynamics of innovation in two Australian industries where creativity and innovation are significant—building construction (for an example of information and knowledge exchange) and textiles and clothing (for an example of power relations).

Information Flows between Players: Innovation in the Building and Construction Industry

The Australian building and construction industry is often considered as weak in innovation. This is largely because the major players or building "assemblers" do not undertake much R&D, an indicator assumed to be the hallmark of an innovative industry. To take this view is to misunderstand the nature of the building and construction industry in which the key players—the "lead firms", companies which build large office or residential towers and undertake complex engineering products—make use of innovations from the supplier firms of products assembled by the lead firms, e.g. steel fasteners, prefabricated walls, etc. The supplier firms conduct the

R&D or collaborate with research and technology organisations to develop the products and processes that eventually underpin the innovative organisational, operational and building practices of the lead firms and their clients. In Australia, lead firms, such as Lend Lease and Leighton or Walker in office construction and Mirvac and Stockland in housing development, operate as the integrators and assemblers of the goods and services needed to construct and equip office buildings, bridges and houses. Their contributions to overall innovation in the building construction sector are critical; if lead firms resist the innovative work of others and do not innovate themselves, the sector stagnates.

The importance of partnership between two lead firms and regulators was seen, for instance, in the design and building of the innovative 2000 Sydney Olympic Village (see Marceau and Cook 2007). One expects creativity in the high-profile knowledge-intensive areas of the industry, especially architecture and interior design. But innovation also occurred in the design and creation of new energy-saving building materials and components, tools and equipment for the construction itself.

The auto and mining sectors and even the PC industry (think Dell) also have this feature of "assembly" of many innovations made by others. Supply firms thus often score highly on R&D investment measures, especially in materials and equipment development, while lead firms innovate in integration and in ensuring that all new elements work together effectively and safely, in the manufacture and construction process and final product.

In the building construction industry innovation and creativity go hand in hand, encouraged by regulators who must ensure construction safety and compliance with urban design requirements, as well as, increasingly, responding to sustainability requirements by regulators and clients.

Understanding what drives innovation and creativity in the building and construction sector requires an understanding of how players relate to each other as well as the more obvious "design" creators. As Gann (1996) has said, it is useful to understand building and construction as a manufacturing process as this focuses on the assembly side and indicates where innovations as well as creativity have their place.

The building and construction industry has several segments, including engineering construction, residential housing and office building.

Engineering Construction

Here we focus on the engineering construction segment of the industry, e.g. bridges, and major buildings such as sports stadiums. This segment is very knowledge- and design-intensive and highly regulated. In our map of the engineering construction segment of the industry the players we identify are: "users", i.e. the lead firms; the producers who produce equipment, materials and services; the regulators; and R&D and education/training. Our information flow map of engineering construction depicts a balanced flow of information between the main players and especially dense flows between large engineering companies and the producers of equipment, materials, services—as well as between the companies and R&D providers, training institutions and public regulators. Engineering construction seems to succeed because the producers provide the lead engineering companies with information and imaginative solutions in design and constructability (creativity), maintain close relationships with clients and suppliers (innovation via customer pull and push), comply with stringent regulatory requirements, have a highly skilled, largely professionally trained workforce and access to R&D to help devise and implement new designs. Our map shows that in complex engineering construction in the Australian building and construction industry, the conditions for continuous creativity and innovation are in place, at least in the major companies.

Residential Building

An information flow map of the house building segment of the Australian building and construction industry shows a very different picture. It is an unbalanced map in which the house building companies are sent information and knowledge by regulators and research and training organisations but do not reciprocate. There is a thin, mostly one-way flow of information from firms producing new products, materials and design to the house building companies. Predictably, as the map leads us to suspect, there is little innovation or creativity in the Australian housing construction industry. The segment is dominated by small and medium-size companies which tend to build more of the same each year with only minor design changes and little thought to sustainability. R&D knowledge available to the housing industry is slow to disseminate, advanced skills are in short supply (and rarely used) and new ideas are seldom pushed by the larger housing developers, especially in regard to single dwellings. Some of the larger and more socially aware developers, such as the NSW state government's Landcom, and local government councils with sustainability policies and concerns about public access send information and push for innovation, but the picture is unimpressive. In this segment of the building and construction industry there is potential for creativity and innovation but it is not realised.

Office Building

There are opportunities for innovation in other segments of the industry, in particular the office building segment, where regulation of energy use is a key to unlocking a wave of innovation. There are players with design skills, knowledge and creativity, but the major companies (both developers and end-users) may need greater incentives to shift direction and shape markets (Malin and Lesson 1995; Marceau and Cook 2007). But that is another map, in which the information flow and power relations dimensions intersect.

Power Relations between Players: Innovation and Creativity in the Textile and Clothing Industry

In my other chapter in this volume (Chapter 4) I give an example of innovation in the Milan garment districts which enabled the Italian textile industry to compete successfully against competition from Asia in the 1980s. Innovation in the Italian garment industry was based on collaborations, information exchange and close proximity of firms.

In this section I describe a map of the Australian textile and clothing industry and its main players depicting the power relations between them. It is a case example of how exercise of power by key players in combination with the opportunities provided by new technology, in particular IT, can produce major transformation in an industry. The Australian story is about the power of large retail stores to push change, and then for small niche boutique designers to respond with their own innovations to revitalise an industry in decline during the 1970s and 1980s. I proposed earlier that, in general, parity in power relations in an industry is more conducive to innovation, consistent with the open innovation system paradigm. Here, paradoxically, we have a case where innovation is driven by a powerful player using the influence that comes with power. However, I suggested earlier that in a time of crisis a powerful player or combination of powerful players can help save an industry by smart application of power.

The essence of the story is that transformation in the textile and clothing industry has been based on a combination of two factors—new technology and shifts in power relations between dominant players, one driver enabling and supporting the other. The new information technology helped provide a platform for the large retailing firms to change organisational and operational arrangements and use their market power, global sweep of product supply chains (e.g. Just-in-time production from overseas suppliers, etc.) to control and influence how the other players—designers, pattern makers, manufacturers, suppliers, wholesalers—would operate and how much they would get for their product or service. The case demonstrates the power of one dominant player to drive change in an industry and in the locus of innovation. The textile and clothing industry is especially interesting because it belongs to the "creative" (fashion) end of the industry spectrum. It is an industry in which creativity meets innovation by the very nature of the product and increasingly the processes used to create the product.

The textile and clothing industry in Australia went from decline in the 1970s and 1980s to transformation and revitalisation from the 1990s. The transformation was accompanied by government regulatory changes (tariff reduction and abolition of protectionist quotas), technological improvements (e.g. laser cutting) and the benefits of government policy attention and assistance. These changes form the background to the transformative changes in the industry over the past twenty years.

Major retailers in Australia in the 1990s became the lead players in the "assembly" and distribution of clothes and other textile-based products. The policies of retailers regarding product supply and management of relationships with suppliers strongly influenced the "creativity" of designers and suppliers. Following the lead of Marks and Spencer in the UK and of supermarket chains in the 1980s, the major retailers in Australian textile and clothing industry acquired a dominant power relationship in dealing with suppliers, wholesalers, designers and manufacturers. The retailers sat on top of the industry, exerting control and influence over the other players.

New technology, especially information technology, helped transform the relationship between suppliers and customers in the product system. An example is the impact of bar-coding and the capacity to order and send product Just-in-time, on supply chain relationships. Bar-coding enabled the retailers to take purchasing control and greater charge of their supply chains. Laser technologies transformed pattern making and material cutting in manufacture of clothing, enabling firms to reduce costs and waste. The new information technologies also became the vehicle for designers to build strong partnership with IT providers, for example, in the use of computer software in the design of new styles and products.

The small design firms, the carriers of product creativity and IP, were also influential players in the industry, but it appears to a far lesser extent than the dominant retail companies. Designers had a strong relationship with providers of IT and computer software and with the assemblers of textile and clothing products, but had limited influence over the other players in the industry, viz. suppliers, contract manufacturers, producers of materials and metallic accessories, providers of machinery and equipment.

There are several lessons about industry product systems and innovation to be learned from the Australian textile and clothing case. First, the use of power/control by a dominant player—in this case large retailers—may force innovation in production methods and design as supplier firms are pushed into new ways of operating. However, there is always the risk that tough requirements by a dominant player might stifle creativity in others or lead the more creative in the industry to rebel. Indeed there has been a "revenge" of creativity in the industry over the past ten years led by talented young designers who have assembled small commercial teams, launched their own labels, opened boutique shops or taken rented space, so as to challenge the dominant power of the large retailers. A second message is the compelling role of new technology as a driver and enabler of smart management and of the high level of creativity involved in design (IT design software) and production. In sum, new technology is a huge part of a story underpinning the innovation that transformed the Australian (and no doubt other countries) clothing and textile industry over the past twenty years.

DISCUSSION AND CONCLUSIONS

This chapter described several approaches to the study of innovation and creativity in the world of industry. Case examples were presented to illustrate where creativity and innovation are located within the dynamics of different industry sectors and their associated product systems. The analysis identified four major players in an industry innovation system: producers, users, R&D/ training institutions and regulators. Several players—R&D/training and regulators—are often invisible in government industry policy considerations.

The case examples illustrate how within an industry the relationships between the major players and lesser players on two dimensions—flow of information and knowledge, power and influence—are the drivers that promote or stifle innovation and creativity in a product system. We saw examples of transformation and innovation (engineering construction; textiles and clothing) as well as an example of weak innovation (housing construction). Application of new technologies is a recurring theme in the examples of industry innovation. We observed from examples the differences in the relations between the key players that help explain differences in the levels of innovation even within sectors (e.g. Australian building construction industry).

I hope that policymakers give recognition to the fact that innovation processes and sources of "inspiration" vary considerably across industries. A focus on the specific characteristics of a particular industry and product system for innovation processes is fundamental to sound policy. Hughes and Grinevich (2007) suggest that a generic approach to industry innovation policy is inefficient, especially when directed at "small and medium" firms, without consideration of their production systems and place in supply chains. However, in Australia recognition of the principle has been slow in the face of a prevailing view that "generic" policies are suitable for industry development, e.g. tax concessions for all R&D. A review of industry innovation policies in several OECD countries (Marceau and Wixted 2003) reveals most innovation-related policies are generic and ignore such factors as type of industry, size and age of firms, degree of international competition and international inputs to the supply chain and whether firms are regularly developing, using or commercialising new technologies. Government policies in those countries took little account of differences between sectors and product systems in the dynamics of innovation.

Among the major components of innovation depicted in Figure 3.1, R&D activity/intensity and protection of IP have been a focus of analysis and policy attention for national innovation. There is now a growing interest in the role of design (the creativity component) in innovation. Design has become a more important element of innovation in many industries as technological advances open new possibilities to enhance the design process as well as production and marketing. Design is also involved in the creation of the "product-service packages" underpinning the competitive strategies of many firms (Marceau, Cook and Dalton 2002).

The creativity involved in design is central to modern economies. Design for mass markets is a critical element in the success of highly innovative furniture companies in Italy and clusters of small firms in Denmark. The creativity involved in design enables firms to make products more robust, more functional or better-looking, to tailor products and services for specific markets, to simplify products, thus enabling firms to service their products more easily, to improve durability of products through use of new materials or components and, by the use of recyclable materials, provide savings to both producers and customers.

Creativity may be separate from innovation in the sense that creativity alone cannot sustain innovation, but must be embedded in suitable organisational, technological and regulatory frameworks if creativity is to become innovation. This apparent paradox also arises in good part from the critical importance of market pull in most innovation success. In reality, creativity and innovation are inseparable. Future studies should examine the links between levels of analysis—national, regional and so on—so that firms and policymakers have a clearer picture of the dynamics of innovation in their fields and the relationships between components of innovation, the organisation of their operations and competitive success. These studies could examine the relationships between successful firms, as in studies of networks and clusters, collaborations, team management and in technology management, the heart of much innovation. This focus would improve the prospects of policymakers to find ways to encourage creativity in addition to building innovation. Advances in knowledge will, however, depend on scholars in both the "innovation" and "creativity" domains recognising the overlap of their subject matter in practice as well as principle. Advances will also depend on the willingness of scholars in different disciplines to link their efforts and take account of insights from other disciplines in the study of creativity and innovation.

NOTES

1. The research literature in the field of innovation is extensive. The writers cited here give a sense of the chronology. The many journals in the field and Handbooks of Innovation are the best sources of recent, empirical work.
2. The case studies discussed in this chapter are based on the work of David Gann, Houghton, myself and others; when at the AEGIS, based at University of Western Sydney, we developed what became known as the "product system" approach.

BIBLIOGRAPHY

Arrow, K. 1962. "Economic welfare and the allocation of resources for invention". In *The rate and direction of inventive activity: Economic and social factors*, ed. R. Nelson, 609–26. Princeton, NJ: Princeton University Press.

Bessant, J., and T. Venables, eds. 2008. *Creating wealth from knowledge: Meeting the innovation challenge*. Cheltenham: Edward Elgar.

Chesbrough, H., W. Vanhaverbeke and J. West. 2006. *Open innovation: Researching a new paradigm*. Oxford: Oxford University Press.

Collins, J., and J. Porras. 1994. *Built to last: Successful habits of visionary companies*. New York: Harper Collins.

Connell, J., G. Edgar, B. Olex, R. Scholl, T. Shulman and R Tietjen. 2001. Troubling successes and good failures: Successful new product development requires five critical factors. *Engineering Management Journal* 13 (4): 35–39.

Dodgson, M., D. Gann and A. Salter. 2005. *Think, play, do: Technology, innovation and organization*. Oxford: Oxford University Press.

Freeman, C. 1982. *The economics of industrial innovation*. London: Pinter.

———. 1987. *Technology policy and economic performance: Lessons from Japan*. London: Pinter.

———. 1991. Networks of Innovators: A synthesis of research issues. *Research Policy* 91: 499–514.

Gallouj, F. 2002. *Innovation in the service economy: The new wealth of nations*. Cheltenham: Edward Elgar.

Gann, D. 1996. Construction as a manufacturing process? Similarities and differences between industrialised housing and car production in Japan. *Construction Management and Economics* 14: 437–50.

Granovetter, M. 1973. The strength of weak ties. *American Journal of Sociology* 78 (6): 1360–1389.

Hollenstein, H. 2003. Innovation modes in the Swiss service sector: A cluster analysis based on firm-level data. *Research Policy* 32: 845–63.

Hughes, A., and V. Grinevich. 2007. *The contribution of services and other sectors to Australian productivity growth 1980–2004*. Sydney: Australian Business Foundation.

Hyland, P., J. Marceau and T. Sloan. 2006. Sources of innovation and ideas in ICT firms in Australia. *Journal of Creativity and Innovation Management* 15 (2): 182–92.

Lam, A. 2005. "Organizational innovation". In *The Oxford handbook of innovation*, ed. J. Fagerberg, D. Mowery and R. Nelson, 126–47. Oxford: Oxford University Press.

Lundvall, B-A., ed. 1992. *National systems of innovation*. London: Pinter.

Malin, R., and N. Lesson. 1995. *A building revolution: How ecology and health concerns are transforming construction*. Washington, DC: World Watch Institute.

Marceau, J. 1999. Networks of innovation, networks of production and networks of marketing. *Creativity and Innovation Management* 8 (1): 20–27.

Marceau, J., and N. Cook. 2007. "Environmentally efficient capacity in the Australian building and construction industry: Lessons from a survey based around the design and building of the Olympic Athletes' Village at Newington". In *Environmental technology management*, ed. D. Annandale, D. Marinova and J. Phillimore, 461–76. Cheltenham: Edward Elgar.

Marceau, J., N. Cook and B. Dalton. 2002. *Selling solutions: Emerging patterns of product-service linkage in the Australian economy*. Sydney: Australian Business Foundation.

Marceau, J., and B. Wixted. 2003. "Innovation policies in selected OECD countries". Report for the Department of Industry, Technology and Resources, Canberra.

Metcalfe, S. 1998. *Evolutionary economics and creative destruction*. London: Routledge

Metcalfe, S., and I. Miles, eds. 2000. *Innovation systems in the service economy*. Dordrecht: Kluwer.

Nelson, R., and S. Winter. 1982. *An evolutionary theory of economics*. Cambridge, MA: Harvard University Press.

OECD. 2002. *Dynamising national innovation systems*. Paris: OECD.

———. 2003. *The sources of economic growth in OECD countries*. Paris: OECD.

Orru, M., N. Woolsey-Biggart and G. Hamilton. 1992. *The economic organisation of East Asian capitalism*. Thousand Oaks, CA: Sage.

Piore, M., and C. Sabel. 1984. *The second industrial divide*. New York: Basic Books.

Porter, M. *The competitve advantage of nations*. New York: Free Press.

Saxenian, A.-L. 1994. *Regional advantage: Culture and competition in Silicon Valley and Route 128*. Cambridge, MA: Harvard University Press.

Schumpeter, J. 1942/1975. *Capitalism, socialism and democracy*. New York: Harper.

Sloan, T., P. Hyland, J. Marceau and J. Kennedy. 2005. "Sources of innovation: An exploratory analysis of four industry sectors in Australia". Proceedings of the 6th International CINet Conference, 4–6 September, Brighton, UK.

Smelser, N., and R. Swedborg, eds. 1994. *Handbook of economic sociology*. Princeton, NJ: Princeton University Press.

Stewart, J. 1994. *The lie of the level playing field*. Melbourne: Text Publishing.

von Hippel, E. 1988. *The sources of innovation*. New York: Oxford University Press.

von Tunzelman, N., and V. Acha. 2005. "Innovation in 'low-tech' industries". In *The Oxford handbook of innovation*, ed. J. Fagerberg, D. Mowery and R. Nelson, 407–32. Oxford: Oxford University Press.

Whitley, R. 2000. The institutional structuring of innovation strategies: Business systems, firm types and patterns of technological change in different market economies. *Organization Studies* 215: 855–86.

4 Space, Place and Innovation

Jane Marceau

INTRODUCTION

In recent decades there has been an interest in what stimulates innovation in the industries and firms located in particular places and spaces. The interest is expressed in examination of the ways in which innovation by the firm is encouraged by the specific characteristics of location, such as the firm's proximity to sources of knowledge, to lead customers and to other firms with complementary or competing interests and markets. In this chapter "space" describes a geographical area, while "place" is what is done within that area that encourages innovation.

Indeed, interest in place or location as a factor in innovation began early in innovation studies, most notably through rediscovery by academics and policymakers of Marshall's (1890/1920) work on industrial districts, followed by studies of what are known as the Italian "industrial districts" (e.g. Berger and Piore 1980). In the 1980s and 1990s, there was an interest in analysing what had enabled Italy's "old" industry districts for knitwear manufacture to remain competitive, innovative and successful despite the down-turn in European textiles and clothing in the face of Asian competition. Much, it seemed, depended on collaboration as well as competition between firms in the same district and the ability of those firms to find in their own locality the new technologies, labour skills and cooperation they needed to develop innovative capacity (Belussi 1989; Belussi and Caldari 2009). This work was extended by Michael Porter (1990) and colleagues in *The Competitive Advantage of Nations*, which observed that innovation and competitiveness owed much to the presence of several characteristics of particular geographic locations or areas. The most successful firms were working with large, leading-edge and demanding clients in the area, connecting in competitive clusters and sharing skills and markets.

Business clusters and innovative activity are of interest for innovation analysts and for policymakers keen to improve the economic prosperity of their towns, cities, districts and regions. The resurgent discipline of economic geography (e.g. Cooke, Levin, Wilson and Kaufman 2001) took the lead in developing the concept of place, precinct, geographic cluster and social networks in industry and regional innovation systems. This work has been followed by economists linking geographic clustering to trade and other policies and by sociologists mapping the features of geographic clusters comprising start-up technology companies, major research universities, engineers and scientists and venture capital investors. An example is Silicon Valley of Northern California, which for fifty years has been at the forefront of innovation, first in computing and electronics, and then in software, the Internet, media and communications.

This chapter provides a glimpse into some of this work on locality and innovation and suggests how it applies to innovation and economic development in key geographic spaces, especially cities. It is about the transformation of spaces into innovative places.

INNOVATION SYSTEMS: NATIONAL, REGIONAL, LOCAL

Before discussing research on place and space in the innovation field, it is important to note the connection between the work of economic and industrial geographers who study place and the work of innovation researchers who study industry and the firm. Innovation in industry can be viewed through several lenses to provide national, regional, local and enterprise perspectives. Each lens tends to look at a different aspect. The starting point is an understanding that innovation is linked to factors external to the firm that provide both a setting as well as encouraging and constraining conditions. The characteristics of these background factors and their interactions influence the levels of innovation found in particular industries and in sets of firms. While there is little analytical work that brings together macro- or meso-level (national, regional or local) innovation systems and factors affecting innovative activity across firms, industry sectors and locations, it is clear that some spaces are more conducive to high levels of innovation than others. Some spaces become innovation places, while others simply remain spaces. (For a discussion of multilevel analysis of creativity and innovation, see Mann, Chapter 15, this volume.)

We can examine the importance of place for promoting innovation by locating it within a set of innovation systems represented as a matrix with two dimensions: the *geographic* dimension: national, regional and local systems of innovation; and the *industry/technology* dimension: industry sector, firm and technology systems of innovation. Cities have their own system of innovation, located on the geographic dimension. However, the matrix suggests that cities of different location and size will display different characteristics as centres of creativity and innovation in good part by virtue of their connections to different industries, firms and technology. Thus the city as place and space is embedded in a larger system. In this system, to give a fanciful example, national government incentives for R&D and provision of regional transport systems and fast railway communications intersect with the emergence of a new industry, such as production of electric cars, encourage firm relocation and provide access to new technology in a coordinated effort to revitalise a city and its level of innovation.

The matrix also suggests that the innovation systems of large cosmopolitan cities influence the complexion of other systems in the matrix as seen, for example, in the influence of New York, Chicago and Los Angeles on the creative vitality of America as a whole. Thus, the creative spark and innovation of a city such as San Francisco is a function of multiple influences stemming from the nation (United States), state (California), region (e.g. proximity to Silicon Valley), together with the innovation systems of particular industries, firms and technology located in or near the city. In brief, place does not stand alone. It is shaped by national, state/province, regional and industry-technological influences. A brief word about each of the systems will provide context to the discussion that follows about the city as a location of innovation. But first, a point about systems themselves. All systems of innovation, whether at the level of nation, region, city or industry sector, are made up of a set of key elements (institutions and organisations) which individually and in interaction determine innovation capacity and performance. There is no authoritative list of components in each system, although there is general agreement about the main components and many of the same components appear in different systems. Human capital built through education and training is a key component of all systems of innovation.

National Innovation Systems

The national innovation system (Freeman 1987) is a starting point for examining the institutions, structures and practices that underpin innovation capacity and improvement of performance. In Chapter 3 (this volume) I list eight components of a national innovation system. The components include legal instruments for regulating ownership and protection of intellectual property rights, work organisation, human capital and education and training systems, availability of venture capital, access to markets and trade, investment in science and technology research and development (R&D). National innovation, analysts say, is determined by the components and also by the quality of the relationships between components; hence the emphasis on social institutions in "national systems of innovation" (see OECD 2002, 2003; see also West, Chapter 2, this volume). Analysts of national systems of innovation point to good fits and/or weaknesses in the relationships between the institutions and in their links to the industrial structures within the national economy. The assumption that nations can improve functioning of their national innovation system by improving the linkages between institutions and firms is built into the analysis.

Regional and Local Levels of Innovation

The economic geography lens on innovation focuses not on national systems but on regional systems. It examines components that affect the innovative capacity of regions and localities, and especially cities. The components include organisational capacity, skill levels, legal frameworks, labour markets and managerial capabilities. There is now a considerable literature on regional systems of innovation (see Cooke et al. 2001). The term "region" is often used loosely to refer to geographical areas vastly different in size, for example, the "Asia-Pacific Region", the autonomous region of Catalonia ("Catalunya") in Spain with a population of about 17.5 million that includes Barcelona, and small regions such as Oresund, which crosses the national boundaries of Denmark and Sweden and has a population of about 3.5 million. Looseness in specifying and differentiating "region" makes it difficult to build a cumulative picture of the "essential" components of regional innovation systems, and indeed the essential components of city innovation systems as they are embedded in regions. In addition, some cities occupy vast areas, e.g. Tokyo, while others are small city-states, e.g. Singapore.

There is, however, considerable evidence that different regional areas have different patterns of innovativeness in firms and industries. One factor that stands out is the benefit of proximity or co-location. Proximity to other firms, markets and clients, and sources of new knowledge, such as research and technology organisations, is useful when firms weigh the risks and rewards of product innovation. Co-location with major clients may be important to retain business and improve innovation by learning from smart customers. Universities located near a regional population of entrepreneurs can become catalysts for local innovation. And tight-knit regional clusters and networks of industries can be important sources of innovation as they encourage collaboration, productive local competition and market development (Braczyk et al. 2004; Cooke 2007).

In some industries geographical proximity is crucial for innovation and profitability. For example, in some industry sectors that depend on Just-in-time operations, proximity between firm and suppliers is crucial. However, some firms with a niche in a highly specialised product can locate far away and still remain innovative and competitive. There are examples of small but highly specialised and innovative Australian firms that are located many miles away from major clients—in one case more than seven hundred miles.

However, geographical proximity is not everything. Many innovation-poor firms and industry sectors can be found in the same geographic location as highly innovative and profitable enterprises. While the characteristics of a place or region matter, ultimately what is critical to innovative and competitive success is the firm's capacity to recognise and benefit from the institutions and resources in its locality. Understanding the dynamics of regions and localities and how they encourage or inhibit innovation is important for policymakers.

Sectoral Systems of Innovation

Some analysts focus their lens on the different industry sectors that help make up an economy and what drives creativity and innovation in those sectors (see Marceau, Chapter 3, this volume). The analysis shows that different industry sectors have characteristic foci for innovation, often driven by the industrial structure of a country and by the location of firms in supply chains inside and outside the country. Therefore, there is an interest in the study of international supply chains in industry sector innovation and in the patterns of relations between small and large firms (Malerba 2004; Wixted 2009). Innovation within key industry sectors may help drive spaces to become innovation places.

Technological Systems of Innovation

The potential for spaces to become innovative places may depend on the availability and adoption of new technologies in manufacturing and services industries where new platform technologies, notably information technology, have led to major transformation. Some writers (e.g. Carlsson 1997) examine technological systems of innovation where innovations spread through related industries and become platforms for further transformation of an industry and where it is located. Hence technology is also a system.

The finding that there are different profiles or patterns of innovation in different industry sectors was ignored for a long time by policymakers; only recently has it been recognised that different industries may require different policies (Hughes and Grinevich 2007). The challenge for policymakers now is to apply what is known about the dynamics of industry sector innovation to the regions, cities, towns and districts where people work and live.

INNOVATION IN URBAN PLACES: THE CITY

The different areas of innovation described earlier come together in the analysis of cities as places of innovation *par excellence*. In developed countries, one or more major cities dominate a region, and cities are at the heart of most regional innovation systems. Cities, moreover, have their own systems of innovation, comprising the same elements as higher-order systems, e.g. the national system, but more concentrated even though many of their relationships (with firms, clients and suppliers) take place outside the city boundaries. Understanding what makes some cities more innovative than others is therefore critical for understanding what makes a nation innovative and successful. The focus on cities is because most people in Western countries live in or near cities and in developing countries there is increasing internal migration toward the city as a place of opportunity.

Innovation, of course, happens in country towns and rural villages, but there is little evidence to uncover the underpinning factors. In addition, proximity of a small town to a larger city may be a prime factor for innovation so it is hard to tease out what drives the innovation.

Some expect urban innovation to be associated with specialist places, but the body of literature on science parks and specialist employment zones yields mixed findings. Some science parks work well as generators of creativity and innovation, such as the radio telecommunications park outside Aalborg in Denmark (see Dalum 1995). The research clusters in Silicon Valley, California, and adjacent to Route 128 in Massachusetts are also examples of success (see Saxenian 1994), but they tend to have special characteristics not found elsewhere and therefore are difficult to duplicate. In addition, many science parks and employment zones turn out to be real estate developments rather than a cluster of firms seeking to establish close links with others.

Cities seen as whole zones or functional regions usually have all of the elements needed for substantial creativity and innovation. The issue is how to harness those elements. Cities, especially large cities, are where it all happens in terms of economic development and innovation. They are the innovation hubs of the nation, producing almost all patents and other measures of new products and processes in business. Cities are the centres of knowledge generation and diffusion and of skilled personnel. They are important transport nodes and house both people and firms. Cities are where much local planning takes place, especially in terms of investment in new or replacement elements of the built environment, special incubator zones or science parks and many of the other supply-side elements considered necessary for innovation by companies and public institutions. It is in cities that the firms which generate economic activity and the public authorities that organise governance interact on a regular basis and where information related to all aspects of innovation is most freely available. Cities are where innovative partnerships between government, business and communities are easier to arrange as the institutions and organisations tend to be smaller and more permeable, promoting faster problem-solving and more room for policy experimentation. Cities are the loci of most major technological breakthroughs and where universities and related organisations develop and impart new knowledge. This is the overall picture of cities and innovation.

But what makes cities innovative? Theories about the sources of innovation in cities abound. A recent volume I edited, *Innovation in the*

City—Innovative Cities (Marceau 2008) brought together a number of suggestions. The next section draws inspiration from ideas proposed in some of those articles.

The People Equation

To many observers, for example, Jane Jacobs (1969), the "people" aspect of city living is critical to creating an innovative place. Some city characteristics, notably size and diversity, are important for innovation which draws upon interactions between diverse people and diverse firms. Diversity is a characteristic of large cities (Wolfe and Bramwell 2008). Diverse people bring a range of skills and experiences to the city. Diverse firms look for new talent and can operate in specialist industries and markets. Cities, especially large ones, act as melting pots where creative and innovative people can find each other and join with entrepreneurs to create new businesses, cross-fertilise ideas and leverage resources. Cities are also places where relative anonymity encourages experimentation with ideas and shifts from traditional ways of doing things. This can be conceived as the beginning of a virtuous circle in which creativity generates successful innovation which then spreads across areas and sectors, drawing others into the action.

Indeed, in the view of Johnson (2008), the simple fact of agglomeration of large numbers of people stimulates innovation because it creates problems of "order", including health and welfare issues, and stimulates the need for solutions. Some of these involve technological innovation as well as organisational innovation and help to create a culture that accepts change and the "disorder" that stimulates creativity. The many needs of people gathered in a relatively small space create challenges where governments need to act. By doing so, they create markets for new products and services—health and hygiene, law and order, transport, education and so on. Crises, Johnson (2008) says, also give cities the push often needed for finding innovative solutions and allow them to take risks unacceptable at other times.

The role of the "creative class" or creative people feature in several analyses of what makes cities innovative, e.g. Florida (2002). Some analysts attach great weight to the people side of the economic development and innovation equation. In this respect, innovation and creativity go hand in hand and together create economic development. Other analysts highlight the attitudes held by creative people, such as tolerance of diversity, which become characteristic of innovative and successful firms and cities. Highly skilled and creative people are attracted to cities with attractive amenity and high incomes. This in turn stimulates further economic progress as it encourages companies to locate their R&D facilities or begin start-ups in localities where there is a supply of and access to creative people. Several cities have made attraction of creative and skilled people a part of their innovation strategy on the basis that creative people stimulate new business and innovation activities.

The notion that "creative people" are the primary source of urban economic development has been questioned by several critics. Storper and Scott (2009), for example, suggest that creative people gather in cities where amenities are already good and jobs available, rather than being the catalyst for a surge of economic development, as suggested by Florida (2002). N. Leon (2008) returns to the issue of creative people in describing how Barcelona is transforming an old industrial area into "buzzy22@Barcelona", a flagship innovative city zone with a mix of work and living arrangements designed to attract an international prestige population. Leon shows it is not enough to attract innovative professionals; ways must be found to embed the newcomers in the local economic arena. This requires sophisticated city governance that caters to the needs of the newcomers and links their personal and economic lives. Chen and Karwan (2008), in their analysis of Shanghai, show that people attracted to Shanghai come with multiple needs which require local social innovations in a way unforeseen by economic development planners. All writers on innovation in cities recognise the need for highly educated people and a skilled workforce whether they are development catalysts or innovation implementers. What is more challenging is *how* people are to be educated and trained in the most effective and socially acceptable manner. In the recent literature on innovation, little attention has been given to education at the secondary level and for the non-standard school students in many large diverse cities (McCarthy and Vickers 2008; see also McWilliam, this volume).

Knowledge and Innovation

The "knowledge economy" is increasingly reliant on the generation and use of formal and informal knowledge as a driver of production. The knowledge can be scientific or relate to knowledge about customer needs, overseas markets or new organisational possibilities. Both formal and informal knowledge are communicated and distributed by people as they work and move around different spaces and locations. Not all innovation-related knowledge is derived from science or other formal sources. Much knowledge is tacit and held informally, gained from local connections and experiences. The diversity of urban populations makes this kind of informal knowledge available in cities, especially cosmopolitan ones, and offers entrepreneurs access to new ideas for products and services, market connections, language skills and other local information.

Cities are centres of knowledge generation and its rapid transmission. Cities are where most universities and research centres are located, where most high-technology firms are established and operate and where the many elements of the information needed for innovation are readily obtainable. While the Internet and other forms of electronic information gathering have overcome the barrier of distance in information exchange, many innovative breakthroughs still depend on physical and organisational proximity.

There is much research on the roles of research generators (universities in particular) and technology transfer in the creation of innovative firms, cities and regions (e.g. Feldman and Bercovitz 2006). The proximity of firms to good universities (usually in cities) is a benefit for innovation. But some firms still search the world for their Ideas pipelines and establish linkages with universities and research institutes in other cities and countries. As "open innovation" becomes more widely used by firms, the importance of proximity to particular sources of knowledge may be diminishing (see Bessant and Venables 2008). It has been suggested that low- and medium-tech firms benefit most from proximity to local universities as they do not have the resources to search for knowledge further afield.

An explanation for the high level of innovation in many cities is the proximity to other innovative firms in the same or complementary fields which make knowledge "spillovers" easier and more frequent. However, the relationship between inter-firm proximity and innovation is complex (see Ibrahim, Hosein Fallah and Riley 2009; Ponds, van Oort and Frenken 2010).

Some cities have thrived on the foundation of specialist knowledge and expertise. Examples include Milan as a centre of design; London as a finance centre, with sub-sectors in the clothing and other design fields; Los Angeles in the entertainment business; and Austin in music. Silicon Valley is perhaps the most famous innovation-business cluster. It has the advantage of proximity to Stanford University and its scientific and engineering expertise, major IT contracts, relatively inexpensive space for start-up firms, movement of people in and out of the area and the attraction and retention of creative knowledge workers. It is not apparent, however, that flourishing "Silicon Valleys" can be built from the ground up by cities and regions attempting to emulate the original. The likely conclusion is that policymakers can encourage existing clusters to grow, but it is tough to build one from scratch.

Governance

Cities are the home of innovative firms and innovative people. But not all cities are innovative in the same way and to a similar degree. Some cities are "in decline" while others are growing in population and economic performance. There are cities where innovation is a desired goal but for various reasons has not led to urban renewal and development. In some cities companies are the drivers of innovation, while in others public authority is the pacesetter for new ideas and development.

There are many puzzles about the drivers of city development and economic innovation. This is relevant to the role that city managers and policymakers can play in encouraging "laggard" cities, regenerating older and economically vulnerable cities and in maintaining and supporting growing urban areas. It is clear that governance arrangements matter greatly in city economic dynamics. Van Winden (2008), for example, describes the challenges of governance of cities in which the objectives are to maximise levels of innovation among local firms and industries but also to innovate in housing and infrastructure development. In Europe, interest in cities as places *for* innovation has been supported and sometimes led by European Union policymakers and leaders who deal directly with cities, firms and government organisations in member countries.

New political and social mechanisms for solving urban problems and taking cities forward are urgently needed. As an example, cities must now deal with the environmental damage caused by their inhabitants, e.g. polluted rivers, loss of habitat, smog and slums, which makes social innovation increasingly important for dealing with technological and associated development issues. Dealing with these major issues demands both innovative social and governance partnerships in the public realm as a basis for new ways of doing things, new transport solutions, energy pricing mechanisms, ways of raising taxes and spending resources and potentially new governance structures for cities. In the private sector, dealing with these problems means incentives to find new technological solutions even though some technologies play an ambiguous role—some, such as petrol engine cars, are part of the problem, while others, such as electric battery cars, can be part of the solution. Cities that put innovation in the public and private sectors to the fore can trigger innovative trajectories and a virtuous circle that creates new clusters of businesses and new industries, attracts creative people and underpins social, cultural and economic benefit for citizens.

Not all cities are the same. Each city must be understood as a separate case, although many common elements can be found when examining growth and decline. But there is little agreement on the reasons behind different city trajectories, successful and unsuccessful, and the role that policy can play in change (Martin and Simmie 2008). Some writers emphasise the role of the public sector, especially for dealing with the problem of "urban order", for seeking to "brand" their cities as "innovative", "scientific" or "creative", for planning innovation in health and welfare, housing and transport and research infrastructure. Other writers look to the private sector for innovation in the city, emphasising business entrepreneurship, the creation of new firms and the expansion of others, sometimes asking what incentives can be offered to innovative firms to stay or move into the city.

The characteristics of cities relating to innovation and economic growth are not simple. Storper and Manville (2006) have pointed to the need to specify what is meant by factors such as "diversity" and "creativity" and the need to study urban choices carefully if policymakers are to make the right choices.

Athey et al. (2008) recognise that innovation is carried out by firms which depend on different aspects of the operations of city life for their innovation capacity. Firms use the city in different ways. Athey and colleagues suggest there are cities in which local hubs are critical to innovative firms and cities where firms depend on local interactions and networks. Accordingly, city managers need to recognise these differences and invest in complementary assets. Underpinning this view is that policymakers must analyse and make judgments about the needs of different firms/industry sectors in their locale.

The most successful cities, it appears, link innovation in city policies for investment and economic growth through entrepreneurship with innovation in governance, administration and policy decisions and their implementation. In many cases, the one depends on the other for maintaining the virtuous economic circle.

CONCLUSION AND POLICY IMPLICATIONS

Finding solutions for the many problems cities face will of necessity encourage innovation. Technological innovation will be coupled with innovation in governance and involvement of citizens and firms in devising change in the ways in which the city is used. This means new ways will have to be found to involve citizens in civic activities as well as personal innovation in their daily lives, e.g. in urban energy and water conservation. But people hold relatively small amounts of the information needed to suggest innovative solutions to quite specialised issues. Perhaps the new technologies can assist in place-based innovation. M-technologies (mobile technologies) are already being trialled in Europe, and their cousin U-infrastructure (ubiquitous infrastructure) has been put into place in Korea (Lee et al. 2008). There are also new technologies capable of revolutionising information gathering, sharing and storage (Dodgson, Gann and Bhardwaj 2010).

Every city has to learn major lessons and to do so it must inevitably improve information gathering and scanning systems and learn from the experiences of others. But cities cannot assume that what works for one will apply elsewhere. Innovation in the public sector, essential for urban innovation, will have to occur on an unprecedented scale and at a speed which institutions and those who hold power within them will find uncomfortable. In its 2006 overview of the dilemmas faced by major cities, the OECD emphasised those relating to governance and government. These include issues of how to relate the disparate agendas of national, regional and city governments to each other's needs and constraints as they struggle to reconcile broad civil society goals and political platforms, such as social inclusion, with the need to invest in the economic engines of innovation and growth that underpin national prosperity, which happens mainly in cities. Linking public and private interests will be a top priority and will need much institutional and organisational creativity. In particular, cities will have to resolve issues over how and when to develop the public–private partnerships needed to resolve funding and regulation issues and ensure that the broad public interests are maintained while business is stimulated to provide the new products and services for economic sustainability.

Policy coordination is of vital importance as cities seek to address broader issues involving innovation. The OECD overview *Competitive Cities* (2006) focused especially on the issue of policy coordination. It is clear that *packages of policies*, not single-focus interventions, are essential to successful urban growth and innovation. The policy issues are complex and cover such matters as priorities for market activity or community amenity, who should deal with the less desirable consequences of economic growth (e.g. dislocation) and the levels of government that should be responsible for costly city infrastructure. Bradford (2005) suggests what is needed are *place policies* and Cooke (2007) emphasises *platforms of policies* are essential for strengthening milieu and entrepreneurship structure, embedding and connectivity.

In most OECD countries a great deal of experimentation is taking place in city innovation and development. London and Gothenburg have developed solutions to transport issues. Monitoring and learning from these experiments is crucial. Unfortunately, there seems little systematic sharing of knowledge between the innovating cities as their projects come to fruition and are evaluated. Jan Annerstedt (personal communication, September 2008) ingeniously suggests a special "innovation place" for thought-leaders from across the world's cities to exchange and test ideas that could influence urban innovation environments. The "innovation place" would also help build cross-national links between hub cities involved in global innovation through overlapping companies, institutions and individuals. The transformation of spaces into innovative places is perhaps the most complex endeavour that policymakers and business people have ever had to deal with. The task must be shared with informed and involved citizens who have a vital stake in the quality and livability of their cities.

BIBLIOGRAPHY

Athey, G., M. Nathan, C. Webber and S. Mahroum. 2008. Cities, systems of innovation and economic development. *Innovation: Management, Policy and Practice* 10 (2–3): 146–55.

Belussi, F. 1989. "Benetton—A case study of corporate strategy for innovation in traditional sectors". In *Technology strategy and the firm: Management and public policy*, ed. M. Dodgson, 116–33. Harlow: Longman.

Belussi, F., and K. Caldari. 2009. At the origin of the Industrial District: Alfred Marshall and the Cambridge School. *Cambridge Journal of Economics* 33 (2): 335–55.

62   Jane Marceau

Berger, S., and M. Piore. 1980. *Dualism and discontinuity in industrial societies.* Cambridge and New York: Cambridge University Press.

Bessant, J., and T. Venables, eds. 2008. *Creating wealth from knowledge: Meeting the innovation challenge.* Cheltenham: Edward Elgar.

Braczyk, H., P. Cooke and M. Heidenreich, eds. 2004. *Regional innovation systems.* Second edition. London and New York: Routledge.

Bradford, N. 2005. *Place-based public policy: Towards a new urban and community agenda for Canada. Research Report F/51.* Ottawa: Canadian Policy Research Network.

Carlsson, B., ed. 1997. *Technological systems and industrial dynamics.* Boston, London and Dordrecht: Kluwer.

Chen, S., and K. Karwan. 2008. Innovative cities in China: Lessons from Pudong New District. *Innovation: Management, Policy and Practice* 10 (2–3): 247–56.

Cooke, P., C. Levin, R. Wilson, and D. Kaufman. 2001. Regional innovation systems, clusters and the knowledge economy. *Industry and corporate change* 10(4): 945–974.

Cooke, P. 2007. Regional Innovation, entrepreneurship and talent systems. *International Journal of Entrepreneurship and Innovation management* 7 (2/3); 117–139.

Dalum, B. 1995. Local and global linkages: The radio-telecommunications cluster in Northern Denmark. *Journal of Industry Studies* 2 (2): 89–109.

Dodgson, M., D. Gann and D. Bhardwaj. 2010. "Technology, innovation and the city". Paper prepared for the DRUID conference, June, London.

Edqvist, C., ed. 1997. *Systems of innovation: Technologies, institutions and organisation.* London: Pinter.

Feldman, M., and J. Bercovitz. 2006. Entrepreneurial universities and technology transfer: A conceptual framework for understanding knowledge-based economic development. *Journal of Technology Transfer* 31: 175–88.

Florida, R. 2002. *The rise of the creative class.* New York: Basic Books.

Freeman, C. 1987. *Technology policy and economic performance: Lessons from Japan.* London: Pinter.

Hughes, A., and V. Grinevich. 2007. *The contribution of services and other sectors to Australian productivity growth 1980–2004.* Sydney: Australian Business Foundation.

Ibrahim, S., M. Hosein Fallah and R. Riley. 2009. Localised sources of knowledge and the effect of knowledge spillovers: An empirical study of inventors in the telecommunications industry. *Journal of Economic Geography* 9 (3): 405–31.

Jacobs, J. 1969. *The economy of cities.* New York: Random House.

Johnson, B. 2008. Cities, systems of innovation and economic development. *Innovation: Management, Policy and Practice* 10 (2–3): 146–55.

Lee, S.-H., T. Yigitcanlar, J.-H. Han and Y.-T. Leem. 2008. Ubiquitous urban infrastructure: infrastructure planning and development in Korea. *Innovation: Management, Policy and Practice* 10 (2–3): 282–92.

Leon, N. 2008. Attract and connect: The 22@Barcelona Innovation District. *Innovation: Management, Policy and Practice* 10 (2–3): 235–46.

Malerba, F., ed. 2004. *Sectoral systems of innovation: Concepts, issues and analyses of six major sectors in Europe.* Cambridge: Cambridge University Press.

Marceau, J., ed. 2008. Innovation and the city—Innovative cities. Special edition of *Innovation: Management, policy and practice* 10 (2–3).

Marceau, J., and B. Wixted 2003. "Innovation policies in selected OECD countries". Report for the Department of Industry, Technology and Resources, Canberra.

Marshall, A. 1890/1920. *The principles of economics.* London: Macmillan.

Martin, R., and J. Simmie. 2008. Path dependence and local innovation systems in city-regions. *Innovation: Management, Policy and Practice* 10 (2–3): 183–96.

McCarthy, F., and M. Vickers. 2008. Digital natives, drop-outs and refugees: Educational challenges for innovative cities. *Innovation: Management, policy and practice* 10 (2–3): 257–268.

OECD. 2002. *Dynamising national innovation systems.* Paris: OECD.

———. 2003. *The sources of economic growth in OECD countries.* Paris: OECD.

———. 2006. *Competitive cities in the global economy.* Paris: OECD.

Ponds, R., F. van Oort and K. Frenken. 2010. Innovation, knowledge spillovers and university–industry collaboration: An extended knowledge production function. *Journal of Economic Geography* 10 (2): 231–55.

Porter, M. 1990. *The competitive advantage of nations.* New York: Macmillan and the Free Press.

Saxenian, A.-L. 1994. *Regional advantage: Culture and competition in Silicon Valley and Route 128.* Cambridge, MA: Harvard University Press.

Storper, M., and M. Manville. 2006. Behaviour, preferences and cities: Urban theory and urban resurgence. *Urban Studies* 43 (8): 1247–74.

Storper, M., and A. Scott. 2009. Rethinking human capital, creativity and urban growth. *Journal of Economic Geography* 9 (2): 147–67.

Van Winden, W. 2008. Urban governance in the knowledge-based economy: Challenges for different city types. *Innovation: Management, policy and practice* 10(2–3): 197–210.

Wixted, B. 2009. *Innovation system frontiers: Cluster networks and global value.* Berlin and Heidelberg: Springer Verlag.

Wolfe, D., and A. Bramwell. 2008. Innovation, creativity and governance: Social dynamics of economic performance in city-regions. *Innovation: Management, Policy and Practice* 10 (2–3): 170–82.

5 Historical Approaches to Creativity and Innovation

Simon Ville

INTRODUCTION

Historians have been interested in innovation *per se* but especially for its contribution to economic growth. This contribution has been widely interpreted through new processes and products but also new ways of organising economic and business activity. Historians have had less to say, however, about creativity than innovation. Interest has largely focused upon the end result of creativity, that is, innovation. This is in large part because of the greater interest in the economic and social consequences of innovation than its origins. In addition, creativity is not easily substantiated through historical evidence since it is not so obviously outcome-based, or as easily documented, as innovation. Nor has much been written about the reverse causality, that is, of innovation upon subsequent creativity. However, increased interest in recent years on the role of human capital in economic progress and the development of knowledge sectors has motivated closer historical consideration of the creative origins of innovation.

In this chapter, I will analyse historical approaches to creativity and innovation. Initially, this will take the form of a broad international comparative perspective and then, more specifically, I will address recent Australian historical experience. This will include a focused look at sources of new technology in the inter-war period. In the final section of the chapter, I will address briefly the policy implications arising from the historical survey.

CREATIVITY, INNOVATION AND THE ECONOMIC DEVELOPMENT OF NATIONS

Historians have laid emphasis on technological innovation as both a shorthand to describe different phases of economic development and as a causal factor in transitions between different epochs. One of the key drivers of a nation's nature and pace of economic development is innovation, particularly through the development of cost-reducing processes, the introduction of new products and services and the development of new ways or organising the activities of firms.

The British "Industrial Revolution" from the late eighteenth century was closely associated with the beginnings of a shift from a cottage system of outworkers using hand tools in cotton manufacture to the deployment of machine tools located in centralised factories (Hudson 2004). Thus, innovation was associated with both questions of spatial location and production technology. In addition, innovation was seen as the key to the explanation for this new industrial age: steam provided the wherewithal to power new machinery and, in turn, the railway system and steam shipping that created national and international markets for the products of the new manufacturing era. The chain effect of the new technology of steam rolled through the middle decades of the nineteenth century—steam's use in railways and shipping motivated new advances in iron, steel and engineering and with it a major stimulus to the European economies (Ville 1990).

The late nineteenth century has been labelled a second industrial revolution—major advances in new, more scientifically based industries, and in different countries, were driving a new expansionary phase: German chemicals, electricity and automobiles should particularly be noted (Pierenkemper and Tilly 2004). American firms carried these advances through into the twentieth century, particularly by extending German technology into organisational and marketing innovations. Automobiles were now mass-produced on assembly lines, sold through specialist dealers offering hire purchase, all of this achieved under new governance structures associated with multidivisional organisations (Chandler 1966).

Moving into the second half of the twentieth century, the types and location of innovation shifted once more and with it economic and industrial hegemony. From the 1950s Japanese firms began to challenge those in Europe and North America particularly through holistic innovation in manufacturing systems, known as lean production; new approaches to labour management; and the development of imaginative forms of inter-firm transacting especially just-in-time contracting (Fruin 1992).

The Diffusion and Transfer of Technology

Besides playing a role in the economic development of individual nations, innovation provides us with a closer understanding of the interaction between the economic rise and decline of nations. "Technological leapfrogging" is the ability of emerging economies to invest in the latest phase of innovations unencumbered by the sunk costs and interdependent requirements of older technologies. This process is made the more compelling where a command structure, normally that of government, provides the leadership for a poor, undeveloped economy to invest in innovation catch-up as was the experience of late nineteenth-century Russia (Gerschenkron 1962). It also requires an effective method of technology transfer. Historians have had much to say about the receptacles and obstacles to technology transfer. David Jeremy (1981), for example, identified the key role of skilled British migrant textile workers in the successful introduction and adaption of the cotton industry in nineteenth-century United States.

Related to leapfrogging and technology transfer is the need to distinguish between originators and users of new technology. Originating firms and nations are the first to absorb its economic benefits and have the trading opportunity to sell the innovation to others. However, recipient users, including late developing nations, avoid the costs of developing the technology and may gain more in terms of spanning developmental gaps from its widespread deployment. Thus, based on a "social savings" calculation, some later developing European nations, such as Spain, appear to have gained more from their railway system than its technological originator, Britain (Ville 1990, 167)

History confirms that the choice and duration of an innovation is often not optimal. Part of the explanation for this lies with human cognitive limits. It is also a function of the interconnections between technological systems as the leapfrogging hypothesis indicates. History provides us with the opportunity to operationalise the concept of path dependency, wherein an initially favourable innovation may continue to operate beyond what is economically optimal. The example is often given of the QWERTY keyboard, designed to minimise key clashes on typewriters but still widely adopted for computer age keyboards (David 2000).

So far, our description of the sweep of history is suggestive of the role of so-called critical, heroic or "macro-inventions", which deals with an essentially new technology, or a cluster of, that constitutes a radical break with the past and has the ability to usher in a phase of renewed economic progress (Mokyr 1990). Where they also generated large positive externalities as a "general purpose technology" (Lipsey, Bekar and Carlaw 1998), their impact was substantial and wide-ranging, affecting both the pace of economic growth and the sources of leadership. Examples of this are thought to include steam power in the mid-nineteenth century, electricity from the end of that century, automobiles in the first half of the twentieth century and information technology in the second half. Long-run economic fluctuations, known as Kondratiev cycles, have been associated with macro-inventions, rising with the diffusion of each new breakthrough and tailing back thereafter. However, within each major historical phase of macro-invention lies many individual micro-inventions, which incrementally improve the original concept and often bring the technology to a "tipping point" whereat major economic breakthroughs are reached. To achieve sustained economic progress, Mokyr argues, an economy, or particular industry, must generate both macro- and micro-inventions. Thus, steam shipping finally dominated the major oceanic routes by the 1880s, after decades of incremental improvements to engine efficiency, with major implications for the efficiency of international trade and the emergence of the first phase of globalisation (O'Rourke and Williamson 1999).

The Institutional Sources of Creativity

A modified view of innovation mutes the centrality of the macro-invention and its spreading effects achieved through externalities. Instead, "innovation is perceived as a broad process, pervasively embedded in many industries" (Bruland 2004, 146). Its embracing nature is not the reverberation from a macro-invention but rather "a general social propensity to innovate" as Bruland (2004, 146) noted of eighteenth-century Britain. This perspective provides us with a powerful link between creativity and innovation as general processes. North's (1993, 16) idea of "mental models" describes society-wide belief systems that help individuals understand and interact with their environment. Mental models evolve gradually over time, their constancy enabling us to make some generalisations about populations over longish periods of time. Thus, some nations may have been more "creative" than others at particular periods of history. If the Industrial Revolution was the creative awakening for Britain, then the Renaissance might have been the same for Italy, and the so-called "Golden Age" of the seventeenth century for the Netherlands. While we tend to associate these particular phases of Italian and Dutch history with the creative visual arts, they were also times of significant practical innovations, note the construction of the Dutch system of canals and the innovative output of the Italian dockyards.

The institutional sources of the creative spur behind the principal phases of innovation highlighted earlier have not gone unrecorded. Creativity has variously been associated with major cultural and intellectual movements, types of educational institutions, the capabilities of firms themselves and the facilitating role of government. The so-called Age of Reason and the "Enlightenment" of seventeenth-century and eighteenth-century England, which were associated with a spirit of rational and critical enquiry into real world phenomena, have been widely viewed as an essential prerequisite to the subsequent "Industrial Revolution". This was seen as fostering an environment of individual observation, inventiveness, and the generation of "useful knowledge" as a public good (Mokyr 2002), epitomised by Watt's realisation of the practical implications of the expansiveness of steam in a boiling kettle.

An emphasis upon more formal scientific and technical training in educational institutions provided a breeding-ground for creativity and experimentation in German industry in the late nineteenth century (Arora, Landau and Rosenberg 1999). American firms of the early twentieth century such as General Electric and Westinghouse developed in-house research laboratories capable of developing a series of related technical advances in engineering and chemicals (Chandler 1990). Likewise, Japanese firms contained notable research capabilities, but also drew upon government organisations and incentives to pursue innovation in fields such as steel and computing (Anchordoguy 1988).

Behavioural patterns and social processes help to provide an understanding of how ideas are shaped. Attitudes to individualism and uncertainty undoubtedly impact on the desire to experiment. Individualism
expressed as a willingness to think and act differently from the mainstream will engender new ideas and approaches. A literature exists that
associates de-familisation, the breakdown of large extended kinship ties,
with the fostering of individualistic enterprise cultures, which includes
a desire to innovate (Macfarlane 1978, 1987). Inventiveness requires a
degree of risk-taking given the likelihood of failure; it additionally represents a desire to mitigate sources of uncertainty through the introduction
of needed innovations. White (1992) and Ville (1998) have both argued
for the importance of risk and uncertainty as an organising principle in
the history of Australia. A desire to mitigate environmental uncertainties
helped to shape business decisions and structures, and related to this is
the fact that much creative thinking and innovative activity was designed
to reduce uncertainty.

Social Capital and Trust-Based Networks

Sociologists, economists and, more recently, historians have begun to analyse the role of trust-based networks in sharing ideas and the flow of information relevant to innovation across organisational divides and geographic
boundaries. At the core of this approach is the concept of social capital,
which analyses the degree of interaction among individuals and between
organisations who trust one another. Such information networks help to
determine the extent, nature and direction of the flow of ideas; although
this is not always optimal since networks can have exclusive as well as
inclusive implications (Maskell 2000; Ogilvie 2003). Geographic contigu-
ity among related industries can foster trust and generate reciprocating
cycles of creativity and innovation as firms provide an innovation response
to a perceived need which in turn motivates new creative opportunities;
such is the Silicon Valley story (Lécuyer 2006).

While social capital can help to bridge institutional and cultural divides,
the concept of "communities of practice" explains how practitioners in the
same field or industry can develop a mutually supportive social environment for the flourishing of new ideas (Wenger 1998). The rise of scientific
and engineering societies in the eighteenth and nineteenth centuries brought
together in a meeting-place the demand for and supply of knowledge in the
form of inventors and researchers, on the one hand, and firms that would
adopt the emergent "useful knowledge" on the other. Many such societies codified their knowledge in published proceedings; for example, from
1860 the *Transactions of the Institute of Naval Architects* in Britain published the latest developments in the rapidly advancing field of shipbuilding
technology. This interaction of inventor and user created reciprocal loops

between creativity and innovation, as the former reacted to insights gleaned from the perspectives and needs of the latter.

The accumulation of large stocks of social capital in Britain by the eighteenth century has been viewed as an important prerequisite for subsequent rapid economic growth (Szereter 2000). British migrants are believed to have transported their social capital tradition with them to the United States and other settler nations including Australia (Greene 2001). Such a view is consistent with Laird's recent thesis that successful entrepreneurs in nineteenth-century United States owed much to their social capital connections (Laird 2006). Godley's study of Jewish immigrant entrepreneurship in New York and London, 1880–1914 (Godley 2001), illustrates the role of trust-based networks carried across geographic boundaries to the process of creativity and innovation. Moreover, from a comparative perspective, it confirms the significance of particular national and cultural environments as Jewish migrants in the United States behaved more entrepreneurially than their otherwise identical counterparts in the United Kingdom. His work forms part of a longer and broader historiography focusing on the cultural determinants of economic development, which includes Weiner's (1981) classic study of the development of an anti-business culture in Britain from the late nineteenth century.

Our examples of the economic impact of displaced groups are replicated through history, and their significance for creativity and innovation are heightened when they bring with them complementary stocks of human capital. Indeed, migrants have often been highly talented, bringing with them knowledge and creativity across many areas of the economy and the arts. In such cases, it is often governmental intolerance of diversity and heterodoxy that has driven out creative sectors of society to the detriment of the domestic economy. Mokyr (2005) has used this insight to trace increased toleration of heterodox ideas by European governments in the three centuries after 1450. Analysing 1185 scientists, he estimates a decline in mobility levels, despite improved transport facilities, as European states, competing for economic advancement, embraced their heterodox creative thinkers.

Governments can go beyond benign tolerance to a more active encouragement of creativity, particularly through mitigating many of the potential sources of market failure. Khan and Sokoloff (2004) have shown how the design of smart patent law in nineteenth-century United States made it easier for less wealthy and well-connected individuals to become inventors than was the case in Europe. Well-defined property rights, the enforcement of patent law and the ability to raise finance through the collateral of a patent were all key features of the American patent system. The effect, therefore, was to foster creative activity more broadly throughout society.

Therefore, understanding the role played by particular institutions, such as social networks, government policy and educational and research organisations, and the form of accepted behaviour (norms) between them and among individuals provides the institutional framework in which innovation has occurred. This "innovation systems" approach has been widely conceptualised and analysed in the contemporary innovation literature but has received little attention from historians (Nelson 1993; Edqvist and McKelvey 2000). History, nonetheless, provides the setting for analysing the evolution of distinctive innovation systems, that is, a combination of elements of continuity—key patterns—moderated by historical experience and change. Such patterns or layers, by setting some distinctive ground rules, have helped to give shape and coherence to a multilayered national framework for innovation at the beginning of the twenty-first century.

INNOVATION IN RESOURCE-BASED ECONOMIES: THE AUSTRALIAN EXPERIENCE

Domestic innovation and its creative spur have been focused on resource-based industries throughout Australian history because of their key role in the economy as a share of output, but particularly their dominance of exports. The share of resources production (agricultural, pastoral and mining) in GDP fluctuated around 25–35 per cent from the mid-nineteenth century until the 1920s, thereafter declining gradually to around 15 per cent by the 1980s (Helliwell 1984, 88). The share of resources in exports fluctuated around 40 to 70 per cent (Pinkstone 1992). Staple theory, which emphasises the stimuli accorded economic modernisation through staple commodity exports, has been widely analysed and discussed in Australian historiography (Pomfret 1981; Fogarty 1985). The advisability of development centred on resource industries has been debated for at least half a century. It has been argued that resource-based development is destined to fail since the "windfall" associated with resource abundance has brought in its wake cognitive, societal, policy and economic constraints on development. In the 1990s Sachs and Warner (1995) formalised this perspective into the "resource curse" hypothesis. Recent work has provided something of a counterbalance by indicating that the curse is not inevitable and by investigating what resource-based economies can do to mitigate it (Ross 1999; de Ferranti et al. 2002).

Nations such as Australia, New Zealand, Norway and Sweden testify to the possibilities for successful resource-based development. One element of the debate is whether resource-based development represents a focus on industries with a low technological capability. As a consequence, this may have contributed to a loss of relative international ranking of GDP per capita over the twentieth century as Australia and similar nations missed out on high-growth industries stimulated by rapid technical progress such as automobiles, aviation, complex chemicals and information technology. Such a view is also consistent with a broader academic and popular debate as to whether manufacturing industries should be the principal foundations of any modern economy. As such, the following research questions might be addressed. Are we correct to view primary industries as a low-innovation sector? Does resource-based development restrain a nation from participating in the rapid change and sequential phases of new technology of the manufacturing and services sectors? Has this form of development created a heavy reliance upon imported technology at the expense of a domestic innovation system? We will address each of these questions in turn.

Innovation in the Primary Industries

Resource-based industries are highly dependent upon the nature and vicissitudes of climate, geology and geography, each of which are highly spatially contingent, often requiring a different response across nations or even sub-national regions. Technology provides a means of moderating these influences.

The natural environment that primary industries have faced in Australia has few parallels in other regions of the world, necessitating domestic solutions to many production problems. Drought, poor soil quality and pestilence emphasised the vulnerability of farming to output vicissitudes that have been marked even for such a highly unpredictable sector. Early innovations in the farming sector, therefore, focused on overcoming development obstacles and mitigating cyclical instability. These included the jump-stump plough, drought and disease-tolerant wheats, fertilisers, merino sheep breeding, dams, artesian wells, wire fencing and nets (Raby 1996). Moreover, regional differences in the environment have been marked, farming processes and products varying, for example, between temperate coastal areas, inland arid locations and sub-tropical regions. In mining, Australia by the late nineteenth century began to play a key international role as one of the major extractors of mineral deposits and one of the principal sources of technical change. In contrast to the proliferation of small-scale operations in Australian farming, mining soon became concentrated in the hands of the leading corporate players who had the resources and motivation to drive innovation. BHP, in particular, has used its technological know-how as a competitive advantage in becoming a resource-seeking multinational, for example, in the operation of coal mines in New Mexico, a large copper mine in Chile and a diamond mine in Canada.

Participation in New Manufacturing Innovation

While Australia has not been a key figure in most of the new high-tech industries of the twentieth century, she has shared in many of the benefits they have brought to producers and consumers. Australians, for example, have been heavy users of air transport and information technology products for both work and leisure. This has particularly included the primary industries—aviation, for example, has facilitated crop spraying, ore prospecting and more generally facilitated communication with remote mining and pastoral settlements. More recently, information technology has improved operational efficiency such as through optimal crop watering and the development of electronic auction sales. Australians, in general, have been amongst the largest users, per capita, of information technology products. As such, they have shared in its benefits, which, particularly over the last decade, have favoured users more than producers due to enormous improvements in efficiency and substantial reductions in price. In particular, information technology has facilitated major productivity improvements in the wholesale and retail trades, construction and finance (Gordon 2000).

Australia has participated in high-tech industries where tradability has been limited by the physical cost of importing or the specific needs of the local market, or where government policy has provided subsidies, tariffs or other forms of support to foster a local industry. A classic example has been the automobile industry, where a series of tariff and exchange incentives facilitated the first entirely Australian-built vehicle in 1948 (Conlon and Perkins 2001, 115–16).

Vertical integration and product diversification by major Australian resource companies have provided opportunities to embrace manufacturing innovation. Capabilities initially established in resource industries were often extended forward into processing and, ultimately, final good production. CSR and BHP are both notable examples of this. CSR's early success in the nineteenth century rested on being the first company to install technologically advanced sugar refining plants on a scale that dramatically lowered costs. By the 1930s, its research laboratories, supported by foreign licences, visits to overseas plants and international joint ventures, led the firm to new downstream products, particularly in the alcohol and chemicals industries. After World War II, related diversification into building materials became the company's focus, including the production of vinyl flooring (1949), insulation and hardboard (1959), particle board (1960) and premixed concrete (1965) (Hutchinson 2001, 109–10). Technical efficiency became the company's watchword. BHP vertically integrated forwards from mining to become the steel industry leader with major plants in Newcastle (1915) and Port Kembla (1935). Subsequently, it diversified into a range of related downstream products, which included steel alloys, hot water systems and tools. Significantly, both companies have now leveraged their technical leadership overseas, CSR in the American building materials industry through Rinker, and BHP-Billiton, now separated from its steel-making capability (BlueScope), in many overseas resource industries as noted earlier.

Imported Technology or a Domestic Innovation System?

International technology transfer has been a key part of the innovation process in Australia, particularly outside the resource-based industries. This has occurred through a variety of channels. Many modern manufacturing industries in Australia are dominated by foreign multinationals who have imported innovations as part of their process of establishment and operation. On other occasions, technology has been transferred as part of a joint venture between a local and a foreign firm. It has been estimated that 83 per cent of the firms responsible for major innovations between 1939 and 1953 had overseas affiliations, while 80 per cent of payments by Australian firms for technical know-how in 1988–1989 went to related foreign enterprises (Hocking 1958, 28–29; Bureau of Industry Economics 1993, 122).

It might be inferred from such a heavy reliance upon foreign technology that Australia has lacked a domestic or national innovation system, with most local inventiveness being restricted to some specific, largely primary, industries. Freeman defines a national innovation system as: "the network of institutions in the public and private sectors whose activities and interactions initiate, import, modify, and diffuse new technologies" (1987, 10). Thus, innovativeness includes activities associated with imported technologies. Perhaps most significant is the modification and adaption of foreign technologies to suit local needs, a process requiring significant creative and inventive energy. A sample of firms in the 1970s revealed that 42 per cent of their research budget was spent on modifying foreign technology (Parry and Watson 1979, 107–9).

Gregory identified four distinctive features of the Australian innovation system, each of which has an ongoing historical resonance: (a) low science and technology expenditure, (b) low private R&D, (c) high government financing and participation in research and (d) high dependence on foreign technology. Consistent with its role in many aspects of the Australian economy, government has served as a major provider of finance and of research organisations. Much of this support has been oriented to the rural sector in recognition of the market failure problems associated with a proliferation of small producers for much of our history. Moreover, it represents a response to unique environmental challenges and the realisation that most of the benefits will be captured locally in commodity markets dominated by Australia (Gregory 1993, 325–29). The CSIRO and its predecessor CSIR is a major public-sector research organisation oriented to the needs of resource industries (Schedvin 1987). Other aspects of a national innovation system that might be emphasised a little more include the role of educational institutions, both vocationally oriented such as Schools of Mines and agricultural colleges, and more broadly based universities as providers of pure and applied research. Agricultural and pastoral societies are a reminder of the role of social and community movements in innovation. Stock and station agents have provided a key network node, connecting farmers with a wider business and research community (Ville 2000, 153–61). Finally, the contribution of domestic corporations has perhaps been understated in place of global companies. Local firms have played a role in negotiating joint ventures with overseas firms, seeking out other sources of knowledge and honing their adaptive capabilities.

RESEARCH FOCUS: THE TECHNOLOGICAL DRIVERS OF
STRUCTURAL CHANGE IN INTERWAR AUSTRALIA[1]

> in 1914 [Australia] could barely arm its expeditionary forces with ri-
> fles, [it] is today able to manufacture locally a sufficient quantity of the
> most modern and complicated weapons from warships to guns. (*Aus-
> tralian Investment Digest*, 15 April 1940, 148)

It has long been assumed that tariff policy drove structural change in
the inter-war Australian economy from rural industries towards manufacturing by providing price protection for infant or inefficient industries
(Benham 1928; Anderson and Garnaut 1987). Investment shifts in favour
of manufacturing, however, may have owed more to exogenous changes
in process and product technology than to the impact of public policy.
Thomas (1988, 271) has argued that "Australia's continued march towards
industrialisation was based not on artificial inducements to produce manufactures, but on lower costs, underwritten by increased efficiency and productivity". Manufacturing not only expanded in size, its technological base
and what it produced changed dramatically. Technology, largely imported
from abroad, was the catalyst for change by creating new products and
reconfiguring cost functions. Its adaptation, industry by industry across the
1920s and 1930s, has been mapped by Mauldon (1938), who shows that
there were marked differences in the rate of what he describes as mechanisation between industries and across time.

The key to this pattern of technological change was the emergence of
two new general purpose technologies, electricity and the automobile.
Their impact was substantial and wide-ranging. Demand for both products increased rapidly from low starting points during the inter-war period.
These technologies transformed many aspects of both consumption and
production. Electricity provided the technological base for a wide range of
new and improved consumer durables. Automobiles constituted a major
new durable in themselves, which, like the many electrical household
products, heralded major and exciting changes in personal lifestyles. On
the production side, the flexibility, controllability, divisibility and speed
of electrical power provided many productivity-enhancing opportunities,
particularly through the spread of electric motors. Automobile production
created the demand directly for many new related industries such as petrol
refining, the manufacture and repair of a wide range of vehicle parts and
the construction of roads and parking stations. Motor vehicles increased
factor mobility across many industries, especially through improved access
to raw materials and better commuting opportunities for workers in labour-
intensive manufacturing.

In both cases, therefore, these new technologies created a clustering of
new industries around them, but also provided productivity improvements
in many older and unrelated sectors. Finally, note should also be made

of the impact of these technological breakthroughs on the service sector, including public transport (trains, trams, buses and taxis), distribution systems (road vehicle transport), retail (store organisation and presentation), finance (vehicle hire purchase) and leisure (moving pictures, holiday accommodation and recorded music), which in turn fed back into further demand for manufacturing products.

A recently constructed database of most new capital issues in this period reveals the acquisition and adaptation of foreign patents by innovative-minded domestic firms to be a central part of this process of industrial transformation. The capital issues information, extracted from the *Australian Investment Digest*, contains evidence of 2,176 new issues across the inter-war period and, when compared with stock exchange data for increased company capitalisation, it appears to have captured most new issues. The aggregate trend for the number of new capital issues and the amount raised over the inter-war period describes a pronounced cycle similar to other measurements of economic fluctuations in Australia. There was a steady rise of capital issues until the onset of the Depression around 1929 when their numbers fell sharply, followed by a recovery from the mid-1930s. The distribution of new issues, either by number or value, across the major economic sectors confirms the conventional wisdom that resources were being shifted into manufacturing.

The question of investment motivation is aided by information on the reason for the capital issue, which has been coded into some standard explanations. Interestingly, more than half (58 per cent) of the value of new issues in manufacturing was derived from new companies. Twenty-five per cent of new issues were by new companies seeking to purchase the rights to manufacture and/or sell another company's products. This was predominantly about acquiring a patent from the inventor or seeking to replicate domestically the success of a product in a foreign market. This was the largest individual motivator and, in the case of new companies, accounted for almost half of the investment decisions. The subdivisions where this purpose was most significant were transport equipment manufacturing, accounting for 52 per cent of its issues, followed by petroleum and coal product manufacturing and polymer product and rubber product manufacturing (each 50 per cent), then machinery and equipment manufacturing (45 per cent). These were the new technology industries of the period most closely associated with electricity and the automobile. Sixty per cent of issues with this purpose (seeking to purchase the rights to manufacture and/or sell another company's goods or services) were directly related to the technologies of electricity or vehicle production.

The figure of 25 per cent understates the significance of innovation since a further 12 per cent of new issues were merely declared as start-up capital for a new company, and a further 18 per cent as expansion or improvement capital for an existing company. A further 20 per cent of all new issues by existing companies did not state a reason, secretiveness doubtless playing a role for some innovating enterprises. Despite the lack of detailed explanation for most existing companies, we expect many purchased new technology through licences and patents and paid for other companies' brands. Thus, an extreme interpretation is that as much as 75 per cent of new capital issues in manufacturing were motivated by a desire to innovate. If innovation is interpreted in the broader sense to include organisational restructuring, the proportion rises above 80 per cent.

The expansion of existing companies was common in more mature industries such as food, beverage and tobacco products; textiles, leather, clothing and footwear; and primary metal and metal products. Even in these mature industries, however, there were a notable number of new firms. In food products, possessing one of the largest shares of capital issues, more than a quarter of issues were made by new companies seeking to purchase rights to another company's goods or services (16 per cent) or seeking start-up capital (13 per cent).

While this research throws light on the role of innovation in structural change, we have yet to discover the origins of the creative spur behind this outpouring of innovation and adaptation. The opportunities provided by the new general purpose technologies undoubtedly motivated a response in Australia as in many other nations. However, the wide range of innovation across industries and firms old and new is suggestive of a broader propensity to innovate, which goes beyond mere imitation of overseas innovation. Australia went through a structural shift from primary to secondary industries that contrasted with the old to new manufacturing industry shift in many other smaller advanced nations in Europe that were importing American and British technology. If Australia's experience was quite different and more marked than most nations, historical landmarks may play a role, particularly the significance of federation, World War I and the broadening of trade routes and migration patterns, in creating an Australia that was more independent and confident of its position in the world and was developing a much wider range of international ties and associations. Put another way, it may well prove to be the case that rapid institutional changes in early twentieth-century Australia lay behind a notably innovative phase of economic development. Comparisons with New Zealand, a resource-based economy that experienced more muted institutional change but also less sectoral diversification, may be instructive.

IMPLICATIONS FOR POLICY

What policy implications, if any, may be drawn from our historical survey of creativity and innovation?

Innovation has come in many forms (product, process, organisational) and is clearly a major driver of phases of transformational economic change and changing industrial leadership among nations. The questions that arise from this statement are pertinent for future policy. In particular, how do nations make the most of the flow-on benefits from phases of innovation—making the right choices among technological alternatives, maximising the positive externalities and optimising its duration.

A range of considerations may influence the choice of technologies at any one time, should the focus be on a nation's areas of comparative advantage or embrace the opportunities for diversification presented by innovation. Australia's approach has manifested various options—strong continued emphasis upon comparative advantages in resource-based industries, but diversification into manufacturing in the inter-war period by adaptation of foreign technologies.

History confirms that general purpose technologies have a powerful transformational role although the principal beneficiaries are not always obvious—continental European gains from the railways and Australian gains from electricity and the automobile. In most of the high-growth innovative industries of the twentieth century Australia has been an adapter and user of technologies developed overseas. As we have seen with ICT over the last fifteen years, there are many benefits from being a user nation. What is critical, however, is the ability to envisage the potential role and application of foreign-derived technologies, the facilitation of its transfer and adaptation and the establishment of incentives for its domestic pervasion.

Finally, optimal duration is about acknowledging that macro-innovations are followed by many years of incremental micro-inventions that transform the efficiency and impact of the original innovation. The ability to gain leadership at the incremental stages can have wide-ranging implications; note, for example, the success of Japanese computer companies following initial leadership by American firms. However, duration is also about regime change—why do problems of path dependency emerge and how does regime change occur among nations and industries? While there has been an historical focus on explaining the rise of British, American and Japanese manufacturing, it would be equally valuable to understand more about the leaders' fall from grace.

Australia presents particular innovation challenges—as a small nation with a comparative advantage in resource-based industries. We suggested earlier that institutional structure is more important than industrial location for a nation. Nor is smallness necessarily a barrier to innovative activity. If innovation leadership remains a possibility, there are two approaches worth pursuing in light of recent historical trends. In the second half of the twentieth century, Australia's population and domestic market grew rapidly. Yet, in many cases, scale economies accelerated more rapidly, meaning the opportunity to compete in many major industries diminished (Forster 1970). However, the raft of changes associated with globalisation and deregulation has enabled smaller economies to compete with increasing effectiveness at the sub-industry level as international specialisation within global industries expands. There is growing evidence to suggest that while manufacturing's share of Australian GDP has been contracting recently, that output is increasingly efficient, competitive and innovative (Anderson 2001, table 13.4).

A much more recent development is the growing global concern for more efficient management of our natural resources for fear of the consequences of depletion and pollution. This throws the emphasis back upon innovation in resource-based industries and the opportunity for nations like Australia to leverage their expertise here. Recent developments in geo-sequestration technology are an example of this. History teaches us that resource-based economies can be highly successful and innovative and that it is the broader question of institutional framework that determines performance, not the sectoral emphasis of production. As Blainey (2006, 11) noted in a recent survey of the history of Australian innovation, "The history of agriculture in the last 150 years is the history of innovation."

If many nations have experienced periods of creative awakening, what can governments do to foster a creative society and economy and to translate a sense of creativity into Mokyr's (2002) "useful knowledge"? Valuing creativity and heterodox thinking is a message that emanates clearly from the historical literature. Investment in human capital may be one response but the solution is also about the learning system itself and how we learn. Tolerating unorthodoxy and pure undirected research and accepting that many areas of creative thinking and research will not produce any tangible outcome are part of the story. So too is fostering a strong sense of trust, cooperation and sharing as reflected in the concept of social capital. The treatment of science as a public good and the interaction of scientific researchers and practitioners in eighteenth-century Britain provide lessons for the twenty-first-century policymakers grappling with the significance of open source technologies and community-style websites as receptacles for shared learning.

NOTES

1. This section is taken from joint research work with Professor David Merrett with the assistance of Mr. Andrew Parnell under ARC Discovery Project 0557412 "Business Profitability and Long Term Industrial Change in Twentieth-Century Australia".

BIBLIOGRAPHY

Anchordoguy, M. 1988. Mastering the market: Japanese government targeting of the computer industry. *International Organisation* 42 (3): 509–43.
Anderson, K. 2001. "Australia in the international economy". In *Reshaping Australia's economy: Growth with equity and sustainability*, ed. J. Nieuwenhuysen, P. Lloyd and M. Mead, 33–49. Cambridge: Cambridge University Press.

Anderson, K., and R. Garnaut. 1987. *Australian protectionism: Extent, causes and effects.* Sydney: Allen and Unwin.

Arora, A., R. Landau and N. Rosenberg. 1999. "Dynamics of comparative advantage in the chemical industry". In *Sources of industrial leadership. Studies of seven industries,* ed. D. C. Mowery and R. R. Nelson, 217–66. Cambridge: Cambridge University Press.

Australian Investment Digest. 1920–40. Sydney: Alex Jobson.

Benham, F. C. 1928. *The prosperity of Australia.* London: P. S. King and Son.

Blainey, G. 2006. "Technology and innovation in Australia's historical development". In *Meeting the challenges—Developing an innovation action agenda, proceedings from the Melbourne Forum,* 10–12. Melbourne: Melbourne Forum.

Bruland, K. 2004. "Industrialisation and technological change". In *The Cambridge economic history of modern Britain. Volume 1 Industrialisation, 1700–1860,* ed. R. Floud and P. Johnson, 117–46. Cambridge: Cambridge University Press.

Bureau of Industry Economics. 1993. *Multinationals and governments: Issues and implications for Australia.* Canberra: Australian Government Publishing Service.

Chandler, A. 1966. *Strategy and structure: Chapters in the history of the industrial enterprise.* Garden City, NY: Anchor.

———. 1990. *Scale and scope: The dynamics of industrial capitalism.* Cambridge: Belknap Press of Harvard University Press.

Conlon, R., and J. Perkins. 2001. *Wheels and deals: The automobile industry in twentieth-century Australia.* Aldershot: Ashgate.

David, Paul A. 2000. "Path dependence, its critics and the quest for 'historical economics'". In *Evolution and path dependence in economic ideas: Past and present,* ed. P. Garrouste and S. Ioannides, 15–40. Cheltenham: Edward Elgar Publishing.

de Ferranti, D., G. E. Perry, D. Lederman and W. F. Maloney. 2002. *From natural resources to the knowledge economy.* Washington, DC: IBRD.

Edqvist, C., and M. McKelvey, eds. 2000. *Systems of innovation: Growth, competitiveness and employment.* Cheltenham: Edward Elgar.

Fleming, G., D. Merrett and S. Ville. 2004. *The big end of town: Big business and corporate leadership in twentieth-century Australia.* Melbourne: Cambridge University Press.

Fogarty, J. 1985. "Staples, super-staples, and the limits of the staple theory: The experiences of Argentina, Australia, and Canada compared". In *Argentina, Australia, and Canada: Studies in comparative development, 1870–1965,* ed. D. C. M. Platt and G. di Tella, 19–36. New York: St. Martin's Press.

Forster, C. 1970. "Economies of scale and Australian manufacturing". In *Australian economic development in the twentieth century,* ed. C. Forster, 123–68. London: Allen and Unwin.

Freeman, C. 1987. *Technology policy and economic performance: Lessons from Japan.* London: Frances Pinter.

Fruin, W. M. 1992. *The Japanese enterprise system: Competitive strategies and cooperative structures.* Oxford: Clarendon Press.

Gerschenkron, A. 1962. *Economical backwardness in historical perspective: A book of essays.* Cambridge, MA: Harvard University Press.

Godley, A. 2001. *Jewish immigrant entrepreneurship in New York and London, 1880–1914: Enterprise and culture.* New York: Palgrave.

Gordon, R. J. 2000. Does the new economy measure up to the great inventions of the past? *Journal of Economic Perspectives* 14 (4): 49–74.

Greene, J. P. 2001. "Social and cultural capital in Colonial British America: A case study". In *Patterns of social capital, stability and change in historical perspective,* ed. R. Rotberg, 153–71. Cambridge: Cambridge University Press.

Gregory, R. G. 1993. "The Australian innovation system'. In *National innovation systems: A comparative analysis*, ed. R. Nelson, 324–52. Oxford: Oxford University Press.

Helliwell, J. F. 1984. "Natural resources and the Australian economy". In *The Australian economy: A view from the north*, ed. R. E. Caves and L. B. Krause, 81–126. Washington, DC: The Brookings Institution.

Hocking, D. M. 1958. Research—the economic implications. *Journal of the Australian Institute of Metals* 3 (1): 23–30.

Hudson, P. 2004. "Industrial organization and structure". In *The Cambridge economic history of Britain. Vol. 1, Industrialisation, 1700–1860*, ed. R. Floud and P. Johnson, 28–56. Cambridge: Cambridge University Press.

Hutchinson, D. 2001. Australian manufacturing business: Entrepreneurship or missed opportunities? *Australian Economic History Review* 41 (2): 103–34.

Jeremy, D. J. 1981.*Transatlantic industrial revolution: The diffusion of textile technologies between Britain and America, 1790–1830s*. Oxford: Blackwell.

Khan, B. Z., and K. L. Sokoloff. 2004. Institutions and democratic invention in 19th-century America: Evidence from "Great Inventors," 1790–1930. *American Economic Review* 94 (2): 395–401.

Laird, P. 2006. *Pull. Networking and success since Benjamin Franklin*. Cambridge, MA: Harvard University Press.

Lécuyer, C. 2006. *Making Silicon Valley: Innovation and the growth of high tech, 1930–1970*. Cambridge, MA: MIT Press.

Lipsey, R., C. Bekar and K. Carlaw. 1998. "What requires explanation?" In *General purpose technologies and economic growth*, ed. E. Helpman, 15–54. Cambridge, MA: MIT Press.

Macfarlane, A. 1978. *The origins of English individualism: The family, property and social transition*. Oxford: Blackwell.

———. 1987. *The culture of capitalism*. Oxford: Basil Blackwell.

Maskell, P. 2000. "Social capital, innovation and competitiveness". In *Social capital. critical perspectives*, ed. S. Baron, J. Field and T. Schuller, 111–23. New York: Oxford University Press.

Mauldon, F. R. E. 1938. *Mechanisation in Australian industries*. Hobart: University of Tasmania.

Mokyr, J. 1990. *The lever of riches, technological creativity and economic progress*. New York: Oxford University Press.

———. 2002. *The gifts of Athena: Historical origins of the knowledge economy*. Princeton, NJ: Princeton University Press.

———. 2005. The intellectual origins of modern economic growth. *Journal of Economic History* 65 (2): 285–351.

Nelson, R, ed. 1993. *National innovation systems. A comparative analysis*. New York and Oxford: Oxford University Press.

North, D. C. 1993. Institutions and credible commitment. *Journal of Institutional and Theoretical Economics* 149 (1): 11–23.

O'Brien, P., ed. 1983. *Railways and the economic development of Western Europe, 1830–1913*. Oxford: Oxford University Press.

Ogilvie, S. 2003. *A bitter living: Women, markets, and social capital in early modern Germany*. New York: Oxford University Press.

O'Rourke, K. H., and J. G. Williamson. 1999. *Globalization and history: The evolution of a nineteenth-century Atlantic economy*. Cambridge, MA: MIT Press.

Parry, T. G., and J. F. Watson 1979. Technology flows and foreign investment in Australian manufacturing. *Australian Economic Papers* 18 (32): 103–18.

Pierenkemper, T., and R. Tilly. 2004. *The German economy during the nineteenth century*. New York and Oxford: Berghahn Books.

Pinkstone, B. 1992. *Global connections: A history of exports and the Australian economy.* Canberra: AGPS Press.

Pomfret, R. 1981. The staple theory as an approach to Canadian and Australian economic development. *Australian Economic History Review* 21 (2): 133–46.

Raby, G. 1996. *Making rural Australia: An economic history of technical and institutional creativity, 1788–1860.* Melbourne: Oxford University Press.

Ross, M. L. 1999. The political economy of the resource curse. *World Politics* 51 (2): 297–322.

Sachs, J. D., and A. M. Warner. 1995. Natural resource abundance and economic growth. *National Bureau of Economic Research Working Paper* 5398 (December): 1–24.

Schedvin, C. B. 1987. *Shaping science and industry: A history of Australia's council for scientific and industrial research, 1926–49.* Sydney: Allen and Unwin.

Szereter, S. 2000. "Social capital, the economy and education in historical perspective". In *Social capital: Critical perspectives*, ed. T. Baron, S. Field and J. Schuller, 56–77. Oxford: Oxford University Press.

Thomas, M. 1988. "Manufacturing and economic recovery in Australia 1932–37". In *Recovery from the Depression: Australia and the world economy in the 1930s*, ed. R. G. Gregory and N. G. Butlin, 246–72. Cambridge: Cambridge University Press.

Ville, S. 1998. Business development in colonial Australia. *Australian Economic History Review* 38 (1): 16–41.

———. 1990. *Transport and the development of the European economy, 1780–1914.* Houndmills: Macmillan.

———. 2000. *The rural entrepreneurs: A history of the stock and station agent industry in Australia and New Zealand.* Melbourne: Cambridge University Press.

Weiner, M. J. 1981. *English culture and the decline of the industrial spirit, 1850–1980.* Cambridge: Cambridge University Press.

Wenger, E. 1998. *Communities of practice: Learning, meaning, and identity.* New York: Cambridge University Press.

White, C. M. 1992. *Mastering risk: Environment, markets and politics in Australian economic history.* Melbourne: Oxford University Press.

6 Economic Approaches to Understanding and Promoting Innovation

Joshua Gans

The economic analysis of innovation has a long tradition and principally deals with how the market and governmental systems provide incentives for economic agents to engage in innovative activity (Arrow 1962; Rosenberg 1982). As with other disciplines, what this means is that economic analysis specialises in understanding one aspect of the process by which innovations are created and diffused within society. As is well known, the creation of an idea is a complex phenomenon and relates to many factors that might include an entrepreneur or scientist's perception of an economic need. However, it is how resources are attracted to the activity of innovation that is the focus of economics. For instance, how do creative people choose to spend their time and how are they supported by other resources (most notably capital) in their endeavours?

In this chapter, I outline the broad economic approach to understanding how resources are allocated to innovative activity. In this regard, I will have less to say about creativity as such although I will make some preliminary remarks in the chapter. My concern here will be with institutions and mechanisms that might encourage the appropriate allocation of resources to innovative activity. To this end, following a discussion of the broad economics of the issue first, I will turn to consider the role of a particular institution—science—in allocating resources, and the implications for public policy in that context. Finally, I explore standard and market-based approaches to the promotion of innovation.

THE RESOURCES FOR INNOVATION

The starting point in economics, for the understanding of how resources are allocated to innovative activity, is to consider what types of resources are desired there. While each individual person and project makes up resources that add to the stock of knowledge, economics tends to conceptualise these in more aggregate returns. This is because it seeks to understand broad patterns rather than what might arise in an individual case.

To see how this works, let A represent the current stock of knowledge (in the world). We let DA represent that addition to that knowledge. The amount of knowledge that is added is a function of various resources. A common representation is this: $DA = f(A, K_A, H_A)$. Here K_A is the stock of capital devoted to knowledge-creating activities and H_A is the amount of human capital engaged in knowledge-creating activities. What this equation says is that all of these factors drive innovation and having more of them (quantitatively or qualitatively) will lead to more innovation. The function, $f(.)$, describes how resources fit together and the productivity of those resources in generating innovations.

Economics has little to say about the nature of the relationship between the actual application of resources and how this leads to the generation of innovations. Rather it is principally concerned about how those resources come to get allocated there. To be sure, that will be related to their productivity in the innovation process. But, critically, this will also be driven by the incentives of agents who control or own those resources and can choose what activities they are employed in.

Incentives to Innovate

Consider an entrepreneur (or any individual or organisation) who has an idea that can be developed into an innovation. To actually generate that innovation (and so add to A), the entrepreneur needs to decide not only whether to allocate his/her own time to this activity but also how to attract any capital that might be required. The case for entrepreneurial attention and capital allocation is similar: if the project is successful, what *rents* (the economic term for profits in this environment) would accrue to those providing resources?

One might think that this might be a fairly straightforward analysis as one might do for any business. But there are wrinkles when it comes to innovation. As Kenneth J. Arrow (1962) pointed out, innovations have specific features that make the case for resource allocation a tough one.

Uncertainty

Let's start with the most obvious: *uncertainty*. There is no guarantee that by allocating resources to innovative activity, you actually produce a useful and commercially viable innovation at the other end of the pipeline. There are technical issues that need to be resolved and, ultimately, it might not be clear whether the end-product is of value to someone. Without either of these, no one will pay for the outcome and so no return will flow to the resources that were deployed to push towards the innovation.

Once again, uncertainty pervades business and economic activity. But innovation creates a different sort of uncertainty—uninsurable uncertainty.

While a business building a power plant can purchase forward contracts to ensure a certain return, the same cannot be said for innovation. If the innovation isn't successful, it is hard to tell if that was because it was a poor idea or because the resources employed did not do their job appropriately. After all, shield the entrepreneur from risk and their incentives to work one-hundred-hour weeks to get the innovation through are diminished. This problem of moral hazard (as it is termed in insurance and agency theory) can mean that no one would want to take the other side of the insurance contract proposed by the innovator. This means that societal means of pooling for risk are not present. And it means that if innovative activity is going to take place and there are going to be those who fail and lose any return on resources allocated, the payoff from success is going to have to cover these (in expectation) as well as the project being rewarded.

Indivisibility

A second issue with innovative activity is that it is *indivisible*. For many economic goods, if you want to produce more of them you put more resources in. For innovation, however, doing that might improve the likelihood of an innovation being generated but, more often than not, unless a certain minimum level of resources is put in place, there is little hope of realising anything of value. You cannot divide up the resources productively.

As an example, consider the task of creating a new Web-based word processing program. At some level, it either works or it doesn't. But if it takes a certain minimum level of resources to ensure that it works, then you had also better expect to make sufficient sales to cover that minimum level.

The profound part of this is the following: in markets, resources are often paid their marginal product (that is, the amount they actually contribute to a project). So if a project generates a certain amount of income, it is possible to ask: if a person put in a little less effort, how much less would that income be? Then you pay that person that amount. The beauty of a divisible world is that you can play this game and in the end you will end up dividing the income between different suppliers perfectly. You neither have bits left over nor are you left short.

When you have some resources who, if they don't supply up to a certain minimum level, nothing gets produced, you have a problem. If you apply the marginal product test to those resources, they should get all of the income. And you can see what happens here: what will be left over for others? So if the market provides others with their marginal product, then those who need to supply a minimum level of effort will be undercompensated. This is a tricky issue when you allow owners of resources to choose what activities they move into. Even where there are non-market mechanisms interacting with these voluntary ones, the adding-up problem still arises. So when it comes to innovation, attracting resources is a particular challenge.

One way to compensate for this is to make up for it at the other end and ensure that the income the innovation generates is high enough. The adding-up issue will still arise but a higher income means there is more left over for those who supply the "minimum effort" resources. I'll note shortly how this becomes a very troubling issue in the case of cumulative innovation (that is, innovations that build upon one another).

Inappropriability

But here we run into the third and final dimension of innovation that distinguishes it from other goods: *inappropriability*. In an ideal world, the limit on the income or the reward from innovation (that is, if we could freely choose it) would be the total value accruing to consumers of the innovation. But, in reality, innovators do not come close to receiving such rewards.

First, at its core, an innovation is an idea. The problem with ideas is that they can be copied, potentially easily and cheaply. This means that if one were to place a new product in the marketplace and that product was observably successful, others could take that idea or something fairly closely related and put competing products in the market. That would be good news for consumers but bad news for the initial innovator. What is more, when weighing whether to direct resources towards innovative activity, agents anticipating such competition and any resulting rent dissipation will likely shy away from those choices.

It is for this reason that the government (in most countries) steps in to protect innovators from such imitative competition. This is what the patent and copyright systems achieve. They offer innovators "breathing space" from potential competition and some assurance they will have time to earn income prior to it being frittered away by competition.

It has been demonstrated that securing patent protection raises imitation costs considerably. Levin et al. (1987) conducted a wide-ranging survey of the research and development policies of companies. To duplicate an unpatented new product, imitation costs were in excess of 50 per cent of the cost of the original innovation across most industries. In some 40 per cent of the industries surveyed imitation costs were 75 per cent or more of original R&D costs. But patenting significantly increased duplication costs in pharmaceuticals (by 40 per cent), chemicals (25 per cent), electronics and semiconductors (7–15 per cent) and machine tools (17 per cent). In very few industries did patent protection prevent duplication of the product in reasonable time.

The problem with intellectual property protection of this kind is that it delays the benefits that might flow to consumers. Most troublingly, the monopoly position of the innovator may lead to some consumers being priced out of the market altogether, as is the case with developing nations and pharmaceutical products.

There has been much written in economics about ways of resolving this issue more cost effectively. One such possibility has been the use of prizes as a reward to innovation. Indeed, the Australian Pharmaceutical Benefits Scheme operates a little like this. When a drug is introduced to Australia, its owner has the option of selling it under a patent in the market or doing a deal with the government. That deal has the pharmaceutical owner agreeing to a lower per unit price for the drug alongside a government subsidy to it. The end result is that the drug is distributed very widely. What is more, because the owner has the option to earn what it would have with an unregulated monopoly outcome, it must be getting more rents from exercising the government option.[1]

A second issue to do with inappropriability is that it can be very difficult to sell an idea free of a final product. Why might an innovator wish to do this? First of all, rather than enter a product market with a whole lot of potential competitors, selling the idea to them can minimise the profit-reducing impact of that competition (for both parties). Second, many innovations do not yield value in isolation but require other complementary assets to complete them. These might be marketing or distribution of resources or even other innovations. Those complementary assets can be built from scratch but they might already exist and be provided by established firms in the industry. Thus, there may be gains to dealing with those firms over commercialisation of the innovation rather than duplicating what they already have (Teece 1987). For each of these reasons, striking a cooperative deal with established firms may secure the innovator more income than taking the product to market themselves.

Faced with these benefits it becomes difficult to imagine why an entrepreneurial firm would not contract with an established firm. One reason is, of course, that the entrepreneur might like to build an empire and keep control over their invention's progress. Another might be that the entrepreneur is far more optimistic than established firms about the invention's value in the marketplace. But even if these were issues, it would still pay for the inventor to *attempt* to negotiate with one or more established firms, if only to see if there are gains from trade despite these other differences. Nonetheless, it remains true that some smaller start-ups avoid approaching incumbents until their products are established in the marketplace.

There is a potentially good economic reason why inventors might avoid negotiations with an established firm altogether: the risk of *disclosure*. In order to sell an idea you have to show the potential purchaser the idea. In some situations a working model is available and its functionality is enough. In other situations, key knowledge has to be disclosed to the potential buyer. The problem is the disclosures themselves may undermine an inventor's ability to contract with unscrupulous buyers. Purchasers may claim they already knew the idea or otherwise fail to reach an agreement. They might then utilise that knowledge to develop competing products and harm the inventor's product market options. Thus, by giving key disclosures, the inventor weakens their own negotiating position to such an extent that it may be preferable to secretly go to the product market and bypass the ideas market altogether (Arrow 1962).

This is not a pie in the sky problem. There are many documented cases of expropriation of intellectual property by established firms. Perhaps the most famous is that of Ford who was found to have violated the intellectual property of Bob Kearns, the inventor of the intermittent windshield wiper. In the 1960s, Kearns solved some long-standing difficulties with developing the wiper, fitted his car with it and drove it down to the Ford motor plant in Detroit to see if they were interested. They inspected the car and employed Kearns for a short time but eventually passed on the idea. In the meantime, Kearns secured a patent only to find later on that Ford and other car manufacturers had employed his technology in millions of vehicles. Kearns eventually won a case against Ford but only after spending twenty years in a legal quagmire. Had he known, he might never have driven the car down to Ford in the first place, let alone develop this important new technology (Seabrook 1994).

The disclosure problem is not insurmountable. One option, of course, is not to disclose the idea. This might dilute the amount legitimate purchasers might be willing to pay to purchase an invention. However, if the inventor could post a bond, they would be able to use this to give a warranty as to the invention's technical and perhaps even commercial viability. This type of assurance requires, at the very least, a well-funded inventor.

A second alternative would be to use a bolder strategy. Bob Kearns might have disclosed his idea to Ford and threatened, quite credibly, to take his invention down the road to GM if they did not strike an appropriate deal. The existence of established, firm competitors can turn the potential expropriation problem on its head, with would-be expropriators becoming the expropriated (Anton and Yao 1994). This type of strategy requires boldness, not generally part of most people's arsenals.

Finally, the inventor could ensure they have sufficient intellectual property protection. Expropriation can occur because of a lack of legal recourse. Having a strong patent or an established trade secret regime allows inventors to disclose their innovations without fear that they could be used against them in the market should a deal not be forthcoming. And it is here that intellectual property serves its additional role. Patent protection not only guards inventors from imitation in product markets but also against expropriation in ideas markets. In some cases, it opens up a new profitable opportunity for entrepreneurs. In other cases, it opens up the only profitable opportunity for them.

The key message is this: while strong IP protection improves returns in both the product market and the ideas market, it makes the latter relatively more attractive than the former. This is because IP protection actually enables ideas markets to work in an economically sensible fashion by insuring the inventor against expropriation. This makes such protection doubly important and, hence, likely to encourage contracting over competition as a commercialisation strategy.

This trend is borne out from a recent survey of start-up commercialisation undertaken by Gans, Hsu and Stern (2002). They found that start-up firms who have strong intellectual property protection (e.g. a patent) are 23 per cent more likely to pursue any type of contracting strategy (licensing, acquisition or alliance) than bringing an invention to market themselves. And this occurred while taking into account differences in the importance of incumbent complementary assets and the nature of the start-up's funding. Moreover, industries where intellectual property protection tended to be strongest (e.g. biotechnology) tended to see more contracting than those where it was weak (e.g. electronic equipment).

The Challenge of Cumulative Innovation

What each of the previous issues—uncertainty, indivisibility and inappropriability—suggest is that the private incentive to innovate, and hence the returns to deploying resources and capital in innovative pursuits, will be well below what we would want socially. If this were a once-off perhaps this dichotomy would not be an issue. But it is because innovation is cumulative that we must worry about under-production.

As noted earlier, the addition to the stock of knowledge is a function of the stock of knowledge itself. Thus, if that knowledge stock is under-produced, future additions to it will continue to lag behind. What is more, because old knowledge generates new knowledge, it is virtually impossible to divide the rents between new and old knowledge producers in a way that can assure maximal innovative activity.

To see this, consider again the role of patent protection. Patents can be set so as to increase the hurdle by which new products must improve upon old products before they infringe upon them. Obviously, increasing the hurdle rate improves incentives for the current innovations but may harm incentives for future ones. Reducing the hurdle rate does the reverse. Indeed, it is very hard to balance incentives so that innovators share and consider the returns to future innovations.

As I will describe in the following, some institutions for innovation—most notably the institution of science—try and get around these problems by looking at non-monetary motivations for cumulative innovation. However, even there, a balancing act needs to be performed.

ECONOMIC APPROACHES TO CREATIVITY

Economics as a field has not dealt with creativity as distinct from innovation.[2] Nonetheless, it is useful to pause here and reflect on how creativity might have an economic meaning, especially in the context of resource allocation.Suppose we envisage a situation where there is a problem to be solved. Innovation can be defined as the result of a problem-solving activity. And in many respects one can imagine that for a given problem and a given solution to that problem there may be a large set of individuals who, with equivalent resources and incentives, could actually reach that solution.

But there are some problems, and in particular some solutions, where we might not be confident about that. An individual may be important and could not easily be substituted for another who would come up with the same solution. These types of individuals and their solutions would be regarded as creative as opposed to innovative. The critical distinction is how translatable the individual is.

This has important implications for resource allocation. For innovation, in general, there is a sense in which it can be planned and created. It just requires the right resources and incentives. But creativity eschews planning. You cannot simply direct resources and incentives and obtain an outcome. Instead, you need to match the individual to the problem, a whole different category of resource allocation.

Economics has done some thinking about matching, but not in this context. This suggests that it may be a good avenue for future research.

ECONOMIC ARGUMENTS FOR SCIENCE

I now turn to consider the economics of a particular mechanism for innovation: science. Examining this will highlight critical public policy choices from an economic perspective as well as showing how the economic approach sits with other disciplinary approaches, most notably, sociology.

The economic argument for the public support of science comes from recognising that science has the elements of being a public good (Romer 1990). First, it is *non-rival* in that it can be widely applied without additional resource cost. Second, it is *partially excludable*. When someone makes use of science they can only appropriate a fraction of its returns (Arrow 1962). However, actions can be taken that allow more or less excludability. These come at a cost of potentially low diffusion and, hence, lower impact on economic productivity.

This is the basis for the public support of science: to correct a market failure that means that private returns from developing scientific knowledge are lower than their social value. This argument, however, does not inform us of the value of supporting science relative to other economic activities and so is an incomplete case. The key idea here is that science is, in fact, a resource allocation mechanism with particular incentives and drivers. It is on that basis that it should be evaluated.

The Case for Science is not the Case for Art

The traditional argument for the public support of science was first developed after World War II by Vannevar Bush, the director of the wartime Office of Scientific Research in the United States. In *Science: The Endless Frontier*, Bush (1945) argued for a distinction between basic research (to enhance understanding) and applied research (to find use). He then stated that basic research was a critical input into the production of useful knowledge (i.e. it is "the pacemaker of technological progress"). This is the *one-dimensional linear model*: science leads to innovation leads to productivity.

In economics this is an argument for spillovers. The idea here is that the public should support science (in its purest form) because this will lead to spillovers on more economically relevant things. The problem arises when it is coupled with another idea in Bush: we need to let scientists be scientists. That is, basic research requires a hands-off approach, allowing scientists to manage their own affairs, allocate resources according to scientific merit and to make a virtue out of being insulated from considerations of immediate use.

In this respect, the case for science starts to look very much like the case for the public support of art. Art leads to "good things" that are unmeasurable and, moreover, require personnel with particular training and experience to understand and appreciate in the first instance. Therefore, good societies support art.

The problem with these arguments, as both scientists and artists consistently discover, is that in times of fiscal restraint, they are not terribly persuasive. And much as they might lament the loss of future returns for short-run expediency, it is a continual issue.

In the case of science, however, this is as much a function of the fact that the basic core argument is incorrect even if the conclusion (long-term funding) is correct. As I will argue here, the problem lies in (a) a misunderstanding of the relationship between science and innovation; and (b) a misunderstanding of the value of pure science as an allocation mechanism for resources.

Use-Driven Basic Research

The premise of the traditional argument for science is incorrect: there is no fine distinction between basic and applied research that makes the former an input into the production of the latter. This case is best articulated by Stokes

(1997). He argues that the notion of the basic scientist is a rare extreme—someone like Bohr and the development of quantum mechanics—as is the applied scientist—someone like Edison. Instead, most research takes place in a form exemplified by Pasteur: it is use-driven basic research. In this realm, the potential for use inspires the quest for fundamental understanding and so the two types of research are fundamentally related.

Once the possibility of use-driven basic research is accepted, the case for the public support of science changes. No longer is it the idea that one puts resources into basic research in the hope of spillovers to applied uses. Instead, the rationale for public support is to encourage research that is driven by use as well as understanding.

Again, this argument itself holds a temptation to another one that itself is incorrect: instead of funding basic research we should stimulate applied research. However, this only reverses the problem by reversing the linearity and supposed direction of causality.

As economists know, a supply-driven argument cannot be separated from a demand-driven one. Hence, a more careful approach is needed to understand why science deserves public support. In particular, one needs to consider science as an allocative mechanism for resources and not merely a "type of knowledge".

Science Is an Allocative Institution

Science is a word that evokes many meanings. However, I want to use it here as a particular way of allocating resources, that is, it is an institution. This view comes from the sociologist Robert K. Merton (1973). To put it in economics terms, science is a way of deciding which projects should be undertaken. First, it is scientist driven in that scientists propose the projects and scientists review them. Second, it has a priority-based reward system whereby there is a *commitment* to give a reward to those scientists who are first to establish a new fact or way of understanding the world. Moreover, these rewards are paid upon success through citation and academic promotion and reputation.

Notice that market-based resource allocations do not operate quite this way. There is more sharing of rewards and the rewards themselves are more immediate: *they stand the current test of the marketplace rather than the test of time.* The latter is believed to be more suitable for sorting out robust facts from currently useful ones.

We do see priorities in the marketplace; however, these are often the result of government regulations. For instance, the patent system rewards priority with monopoly but it is a government regulation. In its absence the rewards, and potentially effort as well, are more diffuse.

Thus, when we talk about public support for science we are really talking about support for it as an institution. That is, do we believe that resources allocated in this manner result in superior longer-term outcomes than pure market-based allocation?

In my opinion, the answer lies in the cumulative nature of knowledge and discovery. This was captured by Newton's famous remark: "If I have seen further [than certain other men] it is by standing upon the shoulders of giants" (letter to Robert Hooke, 15 February 1676). This is the idea that establishing useful scientific facts allows others to build upon that work, leading to more and more understanding. It is an argument for the inter-temporal division of labour in moving the knowledge frontier: that is, that generations of scientists will each specialise in solving a new piece of the puzzle by usefully remembering and working on the work of previous generations.

Fostering cumulative innovation is a more efficient way of solving scientific problems. It allows more specialisation and less duplication of effort. Moreover, it provides a focus on the longer term and necessarily away from immediate priorities.

It is science as an institution that allows inter-temporal externalities to be internalised and cumulative knowledge to flourish. By committing to a reward contingent upon the research's utility in allowing more knowledge to be developed (that is, citation and the judgment of future scientists), and creating a competitive race for those rewards today, better resource allocation is achieved.

In this respect, scientists' concerns for the management of their own affairs and also the continued commitment to future funding (allowing rewards to be realised) are consistent with this view. The case, however, is not paternalistic but practical. Science as an institution evolved to resolve an inter-temporal resource allocation problem, and to continue to be effective it needs to be subject to long-term commitments to resources.

Pure Science and Commercial Research Are Complementary

While science as a resource allocation mechanism does well in resolving the issue of inter-temporal externalities in knowledge accumulation, it does this at a potential cost: a lack of immediate focus on usefulness. Now, as Stokes (1997) points out, scientists are far more focused on immediate usefulness than outside perception; something that even their own rhetoric maintains. This could be for several reasons, not the least of which is that in many scientific disciplines the intrinsic rewards come from seeing knowledge as being useful but also from the pragmatic one that to justify continued funding, it helps to pursue useful knowledge.

This is where science intersects with market-based forms of resource allocation. As noted earlier, the market focuses too much on immediate use and not enough on inter-temporal issues. However, resources for scientific use—other than from the government (but sometimes there too)—come from people with their own immediate needs. The clearest example of this is students. They require knowledge to be able to have productive careers. Academic scientists provide that knowledge and take a proportion of the payment and invest it to continue scientific research. If there is too great a mismatch between that research and immediate usefulness, it becomes hard to supply useful knowledge to students and the funding can dry up.

This happened to the great American universities, Harvard and Yale. They focused on liberal arts education in the nineteenth century and found themselves losing students to universities in Germany that offered more practical-based courses such as commercial science, chemistry and engineering. So the American universities changed tack, introduced new schools and attracted students back.

But similar market-based influence comes from other areas. Business research needs provide funding opportunities for basic research and so scientists wishing to further their careers pay attention to these in their selection of projects. Indeed, it has been increasingly common for academics to accept consulting arrangements with firms for this purpose. To manage that, their research projects are best more closely aligned with business needs. And universities, realising this potential, permit freedom in academic financial arrangements with business. In the end, the confluence of both systems provides for project selection akin to Pasteur, rather than some ivory tower view.

It is the interaction between pure science and commercial research that leads to the selection of research projects designed to not only provide greater immediate use but also address inter-temporal concerns. The latter is supported because of the commitment to the future rewards to scientists. Those scientists who buy into that system are more cost-effective than skilled individuals who do not.[3] Hence, this is a cheaper way of funding research than might be achieved if only commercial considerations mattered.

Thus, pure science and commercial research are complementary systems. Each corrects distortions that would otherwise exist in the other. In the end, it is their interaction that allows use-directed basic research to receive priority. It is this that is the key to longer-run sustained productivity gains in the economy.

A Balanced Approach to Public Support

The complementarity between science as an institution and commercial research implies that public support needs to be balanced in its goals: neither focusing exclusively on immediate use nor on scientific independence.

But more critically it suggests the following: *it is an error to provide public support with the goal of making each system like the other.* To see this, consider first the call whereby public support for research conducted in universities comes with it requirements for its commercialisation, immediate use and intellectual property protection. What this means is that science is being asked to adjust to look more like commercial research. The cost of this is a reduction in the ability of science to function as an institution to reward cumulative knowledge. It raises costs by making scientists more concerned with immediate reward than a future payoff. It also throws up barriers to cumulative knowledge, not only by devaluing it directly, but also by creating intellectual property barriers—the "anti-commons effect".[4]

Similarly, when there is public support given to business R&D, there are conditions tied to this that diminish the value of intellectual property protection and also restrict commercialisation options (say, to be exclusively developed within a home country).[5] These reduce market-based returns to innovation but absent any institution in those firms for cumulative knowledge to do this without a payoff. This is a lose-lose proposition. My concern here is not that the support is unwarranted but that the conditions tied to it undermine the value of the mechanism being funded.

Instead, the government needs to consider providing support, free of restrictive conditions, that allows each system to function as it was supposed to. This means that priority would be given to stable sources of funding for public science with committed future rewards and funding, while for commercial research support should target ongoing subsidies and tax breaks. For each, the government should consider required infrastructure. For science this is to support cumulative knowledge, while for commerce, this is to support commercialisation.[5]

CURRENT WAYS OF PROMOTING INNOVATION

In this section, I review current methods that governments in Australia use to promote innovation and encourage more resources to innovative activity.

Intellectual Property Protection

By far the most common form of government intervention to stimulate innovation is intellectual property (or IP) protection. This protection takes the form of insulating innovators from imitative competition. The benefit of this is that it redresses potential short-falls in the ability of innovators to appropriate profits, whether it be through product markets or through ideas markets. This protection varies from the patent system that prohibits others from commercially exploiting similar products to the innovator's, to copyright, which prohibits unauthorised copying of innovative works, and trade secrets, which allows innovators some control over the flow of information.

However, this method of intervention comes also with a potential cost in terms of the use of the innovation. Put simply, it locks in a monopoly situation for a period of time, leading not only to higher prices but also to difficulties for others in building upon the innovation (that is, it may harm cumulative innovation).

Tax Benefits and Subsidies

A straightforward way of encouraging more resources into innovative activity is a tax credit or a subsidy. Such payments are made on the basis of total expenditures on research and development activities (as in Australia's R&D Tax Credit). Others might be paid based on the employment of scientists or as a tax break on capital investment.

The main issue with such open-ended subsidies is not that they do not encourage innovative activity—they do, and most legitimate expenditures will claim the tax credit. Instead, the problem is that it is difficult to develop easy ways of monitoring whether a given set of expenditures is R&D related or not. Consequently, there is a concern that the subsidy extends across too many activities, costing more than it should for the innovative activity it is encouraging. For the same reason, open-ended subsidies are difficult to budget for and leave governments exposed to fiscal risk.

What is more, in order to claim a tax credit you have to be earning income. Most start-ups do not do this or even intend to do this in the first few years of their lives. Hence, it is a policy that tends to favour big business innovation.

Competitive Reforms

Barriers to innovation and, in particular, appropriability, are regulations and impediments to competitive entry. Government microeconomic reform has been largely directed at "cleaning up" and "freeing up" these aspects of business activity. As Gans and Stern (2003) showed, such reforms were critical in enhancing the innovative capability of Australia in terms of increased productivity brought about by innovative technology adoption and organisational restructuring.

In areas of reform that are ongoing, the potential for innovation policy to support the more rapid seeding of the benefits of such reform should not be ignored. For example, in areas of health, where reforms can see the reduction in waiting lists through more effective patient diagnosis, the adoption of technologies in support of those activities, especially in the public sector, should be given priority. Put simply, the reform "payoff" is likely to be delayed without recognising the need to support private incentives to adopt key innovations.

Grants

Another way of lowering the costs of innovative activity is by the use of direct contributions or grants. These differ from tax credits and subsidies in that they are targeted and not open-ended. Clearly, by providing capital directly to make up for short-falls in expected rates of return, innovative activity can be stimulated.

Grants are not, however, without potential problems. First, there are information problems that arise because scientists know more about the prospects flowing from a project than do funders. They may have incentives to over-claim potential contributions to attract funding, in particular, about potential feasible use as opposed to scientific merit. For example, Kremer and Zwane (2003) report that in relation to agricultural innovation grants:

> Many technologies developed by push program scientists have been adopted at low rates in developing countries because scientists have failed to develop products that address constraints faced by farmers. Advances worthy of scientific acclaim, such as improved cowpeas that defoliate, that have seemed promising in a controlled environment have not translated well to mixed cropping environment in which farmers actually work. (14)

This suggests that while grants can be useful in stimulating innovative activity, there may still be issues in encouraging follow-on investment that yields commercialisable products.

Another issue with grants is that project selection is difficult. In many countries, political pressures can overcome scientific and commercial merit. They involve ties that, indeed, harm continued commercial development, such as requirements that products be manufactured in their home country (see Gans 1998). In addition, political constraints may lead to poor location decisions for innovative activities and sites and also constraints on the ability to attract quality researchers at publicly capped wages.

Given these considerations, it is safe to say that grant programs are currently most effective for more basic research. Indeed, they have been successfully applied in university-based scientific environments. It is useful to consider why they have been more successful there.

Put simply, academic-based research can minimise the costs associated with grants. The reason for this is that granting agencies have an instrument to ensure performance: they can cut off future funding. Combine this with the career incentives of academics and you have the elements of a solid performance system.

To see this, consider the role of a funding agency. It solicits proposals from researchers but it also requires researchers to submit their track record (this is typical in the Australian Research Council, for example). They then adopt rules that deny funding to researchers who have not performed on previous grants. Thus, those researchers have strong incentives to perform, truthfully reveal expectations of success and merit and other desirable things to make sure they can meet those expectations and not be excluded from future funding.

What is also true about this is that it creates incentives to self-select the most productive researchers. A productive researcher knows that they will be proposing many more projects to the agency than unproductive ones. Hence, they have the most incentives to manage expectations. Less productive researchers have little incentives and so will over-claim (and perhaps get away with this once). If the funding agency can also ramp up grants based on an extended track record, this aspect of the system becomes even stronger.

Note, however, that for this to work it requires long-lived funding agencies. Pushes or one-time expenditures on granted projects do not create proper incentives. The funding agency needs to have a memory longer than most political institutions.

MOVING TO A MARKET-BASED APPROACH

In response to issues associated with common approaches to promoting innovation, some have recently suggested more market-oriented approaches.

The pressure for this has come from those who are concerned that current approaches, while promoting innovation, do not sufficiently promote useful innovation and its adoption. For instance, Kremer and Zwane write:

> In an effort to improve adoption rates, recent research programs, such as the Cassava Biotechnology Network, have attempted to identify attractive technological advances by interviewing farmers about their perceived needs. While this may be an improvement over research programs that have no input from farmers' responses to survey questions may depend on how questions are asked; farmers may not know the scientific opportunities and challenges, and there may be opportunities for scientists to manipulate or ignore farmers' responses. (2003, 14)

In that situation, it was a case of simply asking potential users to articulate their needs before presuming that a potential technological advance might satisfy a need.

The newer approaches have come under the rubric of "pull" programs that specify needs first, in contrast to previous "push" programs that specify potential solutions first.

> One of the biggest advantages of "push" programs relative to "pull programs" (other than patents) is that they do not require specifying the output ahead of time. A "pull" program could not have been used to encourage the development of the Post-It Note® or the graphical user interface, because these products could not have been adequately described before they were invented. (Kremer 2001, 40)

In situations where it may be possible to specify and communicate a need, then "pull" programs might be effective in procuring high-quality solutions to key problems that industries face. Put simply, the efficacy in "pull" programs comes from a realisation that there is a potential for the government to facilitate matches between need and technical knowledge and that it is the effort in articulating needs and the search for research that might resolve those needs that is holding effective innovation back.

Here I outline some of the ways in which "pull" programs might be implemented and discuss potential issues or difficulties with each approach.

Prizes

Perhaps the simplest "pull" program works as follows: there is a need with a performance metric. The first research team to resolve the need and exceed that metric gets a cash payment. This is referred to as a "prize" model.

There are examples of such models being sporadically tried by governments and private benefactors (most notably Google's Larry Page, the Virgin Earth Challenge and the X Prize Foundation). More recently, the US DVD rental distribution company, Netflix, has offered a million-dollar prize for a better customer information assessment system that outperforms its existing system by a certain amount (Leonhardt 2007). The contest will run for five years.

Some economists favour prizes over IP protection. Joseph Stiglitz (2007) makes the case for prizes rather than patents:

> A scientific panel could establish a set of priorities by assessing the number of people affected and the impact on mortality, morbidity, and productivity. Once the discovery is made, it would be licensed.

Stiglitz notes the problem with patents for medical drugs and technologies: high price = more illness/deaths. His solution is to ensure low prices by having governments fund prizes for proven innovations. So consumers pay indirectly through the tax system.

Prizes tie the innovative reward to performance and they allow for competition in claiming that reward. It is a highly incentivised system that doesn't actually require any money up front or for failure. However, in order to work the prize amount has to be high enough (there is risk being borne by entrants) and the terms have to be clear enough. What is more, the prize needs to be a credible contract as performance may take years to achieve.

So when it works, a prize system can be highly desirable. The issue is that it doesn't work for all innovations. First of all, many attempts have been made to use the Internet to set up markets whereby firms specify needs and a prize and others compete for them. However, these appear to have been rather thin markets. Second, it is sometimes hard to define performance metrics that can be immediately monitored. For instance, a prize for a vaccine might be awarded only to discover years later that the solution has unwanted side effects.

Related to straight-out prizes are research tournaments. A tournament has a fixed duration and specifies a prize to the research team that makes the most "progress" according to a performance metric. Tournaments are less risky for entrants because a prize will be awarded and there is no risk in terms of a threshold performance target being unrealistic. However, the other side of this is that the promoter may have to pay for whatever progress there is in any case and may not get value for money. In addition, this suffers from many other issues associated with performance metrics, including the potential for politics and favouritism in award decisions. It also runs the risk of collusion amongst participants who might keep effort down. Nonetheless, the Netflix example appears to satisfy this as it is not an award for the first team to achieve a 10 per cent improvement but the team that progresses the most beyond 10 per cent in five years. That might end up buying Netflix lots more than it expected.

Finally, as noted earlier, microeconomic reforms have raised the returns to entry into certain industries. In particular, there are profits to be had to those who access markets quickly and effectively. One can liken this to a prize released by government for innovative entry. This has been most clearly seen in utilities industries with rushes to invest in modern infrastructure as well as in support of other government initiatives (such as the environment), but also in public firms, such as Australia Post, who have faced threats as a result of open access to their traditional lines of business. Finally, where governments have engaged in open tender processes, these have prize-like qualities that stimulate innovation. An excellent example of this is Transurban's development of the eTag system, largely in response to Victorian government stimulus.

Matching Grants

Another form of "pull" policies involves matching grants and is commonly applied in Australia. For example, the Australian Research Council has linkage grants that require researchers to have an industry partner putting in cash or in-kind resources.

The idea of a matching grant is that it requires someone other than the public provider to find value in the research proposal. Presumably, that value arises because of an immediate need as opposed to purely scientific merit. However, its effectiveness rests on the ability of that private agent to assess the quality of those institutions and labs with whom it is linking.

Matching grants are also useful in saving the public funder from having to investigate the overall likelihood of the project's success. It can be presumed that a private funder would engage in such investigations. Hence, so long as the matching requirement is stringent enough, poor projects will be weeded out and more good projects will be able to go ahead.

Advance Purchase Commitments

It was noted earlier that two of the main constraints on private investment in innovation are, first, that monopoly pricing may lead to under-supply of the innovation and, second, that matching issues require investments by users as well as innovators. Michael Kremer (2001) has proposed the use of "advance purchase commitments" to overcome these issues in markets for vaccines.

His proposal is as follows: A government specifies what it is willing to pay per dosage for a vaccine. It writes this as a binding contract. Pharmaceutical companies then compete to, first, develop a vaccine, and, second, if there are competing vaccines, to encourage use of them. For each treatment they sell, they receive a co-payment from the government (perhaps 100 per cent) in return for a fixed price to consumers. In this way, the government can set, in advance, a price based on the social value of the vaccine but it only has to pay this if a vaccine is developed. As noted earlier, the Australian PBS has some of these elements but instead of a prior commitment, it allows pharmaceutical companies the option of monopoly pricing. Kremer's proposal goes further, especially for innovations that would not otherwise be developed.

This proposal allows vaccines to be distributed widely, forces needs to be exposed, is cost-effective (pay only on receipt) and encourages competition. What is more is that there is no reason why it could not apply to innovations beyond vaccines. The main requirement is that the government or industry be able to write an upfront contract to commit to purchases. While this might not hold for basic research, for much applied research it is surely possible.

Creating a market for an innovation is similar to direct procurement. However, with procurement, the tender process identifies the supplier. With an advance purchase commitment, the procurement terms are set and the competition to create the innovation replaces the tender process. Other than that the economic considerations are very similar.

Put simply, it "might be more efficient to have problems seeking solutions than solutions seeking problems" (Hellmann 2007, 626). As an MIT engineer noted:

> With university technologies you pull the technology out and you run around saying "Where can it stick?" It's probably better to say I've heard about these problems and I think I can solve it. But with companies coming out of MIT, it's always the same thing, what do I do with it to shoehorn it back into industry. (Shane 2004, 204)

Thus, the creation of markets for innovations can turn the tables on how matches are realised.

There are three areas where advance purchase commitments would likely be most fruitful initially.

- **Dissemination of existing technologies.** Previous government expenditures have encouraged basic research and also the identification of potentially commercialisable products. An advance purchase commitment could be made for some of these products in order to encourage them to be commercialised and brought to market. In that way, the existing pool of developed technologies could be identified and exploited.

- **Technologies related to government policy.** Government policies—especially where they are areas of active reform such as health, education and the environment—often have innovative needs that are related to those reforms. As part of the reform agenda, advance purchase commitments could be set up.
- **Supporting grant programs.** The government may wish to consider a broad-based grants program for small business innovation (such as the SBIR program in the United States), but add to it an advance purchase commitment to ensure useful results. This would provide some commercial certainty of the new solution, underwrite some initial sales (thereby enabling attraction of private-sector capital, getting back to our discussion of "pulling" private capital into the innovation space) and enhance the development of skills/know-how.

These areas demonstrate a role for government in creating markets for innovations. In addition, government could facilitate industry groups to match funds for this purchase and to associate to identify common needs, in this way internalising spillovers between them.

NOTES

1. Michael Kremer has proposed that a similar scheme be set up to encourage more vaccine development on a global scale.
2. Stern (2004) demonstrates this for scientists. See also Aghion, Dewatripont and Stein (2009).
3. Murray and Stern (2007) show a mild anti-commons effect.
4. Gans (1998) articulates this argument in more detail.
5. Stern (2005) and Duncan et al. (2004).

BIBLIOGRAPHY

Aghion, P., M Dewatripont and J. Stein. 2009. Academic freedom, private-sector focus and the process of innovation. *RAND Journal of Economics* 39 (3): 617–35.
Anton, J. J., and D. A. Yao. 1994. Expropriation and inventions: Appropriable rents in the absence of property rights. *American Economic Review* 84 (1): 190–209.
Arrow, K. J. 1962. "Economics of welfare and the allocation of resources for invention". In *The rate and direction of inventive activity*, ed. R. Nelson, 609–25. Princeton, NJ: Princeton University Press.
Bush, Vannevar. 1945. *Science: The endless frontier.* Washington, DC: Government Printing Office.

Duncan, M., Andrew Leigh, David Madden and Peter Tynan. 2004. *Imagining Australia: Ideas for our future.* Crows Nest: Allen and Unwin.

Feinstein, J. 2006. *The nature of creative development.* Stanford, CA: Stanford University Press.

Gans, J. S. 1998. *Driving the hard bargain for Australian R&D.* Prometheus 16 (1): 47–56.

Gans, J. S., D. Hsu and S. Stern. 2002. When does start-up innovation spur the gale of creative destruction? *RAND Journal of Economics* 33 (4): 571–86.

Gans, J. S., and S. Stern. 2003. "Assessing Australia's innovative capacity in the 21st century". *IPRIA Working Paper.* www.ipria.org.

Hellmann, Thomas. 2007. The role of patents for bridging the science to market gap. *Journal of Economic Behavior and Organization* 62: 624–47.

Kremer, M. 2001. Creating markets for new vaccines. Part I: Rationale, Part II: Design issues. *Innovation Policy and the Economy* 1: 35–118.

Kremer, M., and A. Zwane. 2003. "Encouraging technical progress in tropical agriculture". Mimeo, Harvard. http://www.economics.harvard.edu/faculty/kremer/files/Encouraging_Technical_Progress.pdf.

Leonhardt, D. 2007. You want innovation? Offer a prize. *New York Times,* 31 January.

Levin, R., A. Kelvorick, R. Nelson, and S. Winter. 1987. Appropriating the returns from industrial research and development. *Brookings Papers on Economic Activity* 3: 783–831.

Merton, R. K. 1973. *The sociology of science.* Chicago: Chicago University Press.

Murray, F., and S. Stern. 2007. Do formal intellectual property rights hinder the free flow of scientific knowledge? An empirical test of the anti-commons hypothesis. *Journal of Economic Behavior and Organization* 63 (4): 648–87.

Romer, Paul. 1990. Endogenous technological change. *Journal of Political Economy* 98: S71–S102.

Rosenberg, Nathan. 1982. *Inside the black box: Technology and economics.* Cambridge: Cambridge University Press.

Seabrook, J. 1994. The flash of genius. *New Yorker,* 11 January, 38–52.

Shane, S. 2004. *Academic entrepreneurship: University spin-offs and wealth creation.* Northampton, MA: Aldershot Edward Elgar.

Stern, Scott. 2004. Do scientists pay to be scientists? *Management Science* 50 (6): 835–53.

————. 2005. *Biological resource centres: Knowledge hubs for the life sciences.* Washington, DC: Brookings.

Stiglitz, J. E. 2007. Prizes not patents. *Project Syndicate.* http://www.project-syndicate.org/commentary/stiglitz81/English.

Stokes, D. 1997. *Pasteur's quadrant: Basic science and technological innovation.* Washington, DC: Brookings.

Teece, D. J. 1987. "Profiting from technological innovation: Implications for integration, collaboration, licensing, and public policy". In *The competitive challenge: Strategies for industrial innovation and renewal,* ed. D. J. Teece, 185–220. Cambridge, MA: Ballinger.

7 Creativity and Innovation
A Legal Perspective

Andrew Christie

INTRODUCTION

Put in simple terms, law is a social science discipline concerned with the body of rules, whether proceeding from formal enactment or from custom, that a particular state or community recognises as binding on its members or subjects. Typically, these rules regulate the conduct of the state with respect to its members, and the conduct of the members with respect to the state and to each other. The conduct so regulated is extremely wide, and includes the production and exploitation of the products of the human intellect ("intellectual property"). Specific rules regulating the production and exploitation of intellectual property have been part of the laws of nations for many centuries. Rules governing exclusive entitlements to inventions, for example, have existed in England since the mid-sixteenth century (Dent 2006). By the end of the twentieth century, almost every nation-state in the world had detailed intellectual property laws of common form and content, governed by international treaties.

Given the long tradition of laws regulating the creation and exploitation of intellectual property, it might be assumed that the topic of creativity and innovation would be a well-developed field within the discipline of law. In fact, this is not the case. The law's perspective on creativity and innovation has not been the subject of any substantial consideration to date. This is demonstrated by the fact that a search undertaken in early 2010 of the LexisNexis database, the world's largest database of legal materials, identified only four works with the words "creativity" and "innovation" in the title; and of these, only two (Sawyer 2008; Long 2009) deal in any meaningful way with the topic of the law's perspective on the concepts of creativity and innovation.

This chapter will begin to fill the lacuna of the law's perspective on creativity and innovation in the following manner. It will start by ascertaining what meaning the law gives to the concepts of creativity and innovation and what relevance those concepts have to law. It will then explain the fundamental tenets of the only area of law—intellectual property law—in which those concepts have relevance. Next, the precise role of those concepts will be identified and elaborated. The chapter will conclude with an elucidation of the policy implications of the law's perspective on creativity and innovation.

CREATIVITY AND INNOVATION IN LAW

Legal Meaning of Creativity and Innovation

Neither creativity nor innovation is a concept with a specific legal meaning. Authoritative legal dictionaries, for example, although listing the general legal definitions of many thousands of words and phrases provide no entries for "creativity" and "innovation" (Martin and Law 2006; Hay 2007). It is a fundamental principle of law that, where a word does not have a specific legal meaning, the meaning to be given to it in law is its "ordinary" meaning. The ordinary meaning of a word is evidenced by the definition of it in a dictionary. Thus, from the law's perspective, the meaning to be given to the terms "creativity" and "innovation" are the dictionary meanings of those terms.

According to the *Oxford English Dictionary*, "creativity" means "creative power or faculty; ability to create", and "create" means "to bring into being, cause to exist". It follows that *creativity* is the ability to bring something into being. The same dictionary defines "innovation" to mean "the action of innovating; the introduction of novelties", and "innovate" to mean "to bring in (something new) for the first time". Accordingly, *innovation* means the act of bringing into being something that is new. Similar definitions are found in major dictionaries in the United States of America (*Merriam-Webster Dictionary*), Canada (*Canadian Oxford Dictionary*) and Australia (*Macquarie Dictionary*). Thus, the ordinary—and, hence, the legal—meanings of these two concepts are the same across the major common-law legal systems of the world.

Relationship in Law between Creativity and Innovation

It will be appreciated that there is a subtle, but very important, difference between creativity and innovation as those concepts are understood in law. This is best identified by considering the products resulting from the exercise of creativity (the ability to bring something into being) and of innovation (the act of bringing into being something that is new). The product resulting from innovation—that is, *an* innovation—is, by definition, something brought into being that has not previously existed. In contrast, the product resulting from creativity—that is, a creation—is merely something brought into being, whether or not that something has previously existed. It follows that innovation is a subset of creation; that is, all innovations are creations, but not all creations are innovations. The proof of this relationship is as follows. All innovations are creations because all innovations have been brought into being. Not all creations are innovations, however, because not all creations are things *newly* brought into being. Some thing may be brought into being that already existed. Such a thing, although not new (and hence not an innovation), is nevertheless a creation.

Having considered the relationship between creation and innovation, it is possible to identify the relationship between creativity and innovation. Innovation (in the sense of the *action of producing* an innovation) is a subset of the actions that result from creativity. This is because all acts of innovation result from creativity, but not all actions resulting from creativity are innovations.

Relevance of Creativity and Innovation to Law

Neither creativity nor innovation is a concept that has any particular relevance within the general discipline of law. This reflects the fact that there is no recognised field of law that directly regulates creativity or innovation; that is to say, there is no law *of* creativity or law *of* innovation. Furthermore, with only one exception, there is no appearance of these concepts within the major fields of law. The only field of law in which creativity and innovation have any relevance is the field regulating the production and exploitation of intellectual property. The nature of this field of law and the key characteristics of the legal regimes within this field are discussed in the following.

LAW AND INTELLECTUAL PROPERTY

Nature of Intellectual Property

An informal way of defining "intellectual property" is "all those things which emanate from the exercise of the human brain, such as ideas, inventions, poems, designs, microcomputers and Mickey Mouse" (Phillips and Firth 1995, 3). This classification is consistent with the notion that the subject matters constituting intellectual property are primarily derived from human intellectual activity—hence the word "intellectual" in the title.

Intellectual property is one type of intangible subject matter recognised by the law (another such type is a monetary debt owed by one person to another). An intangible subject matter is one that, although existing, can't be touched. Intangible subject matters may be contrasted with tangible subject matters. Tangible subject matters are ones that can be touched, such as chattels and land—that is, "physical property".

Intellectual property is similar to physical property, in that exclusive rights may be granted in relation to it, and those rights may be transferred by the owner to other persons. However, intellectual property is dissimilar to physical property in three important ways: intellectual property is non-rivalrous, inexhaustible and non-excludable. *Rivalrousness* is a characteristic of physical property, whereby only one person can use the item at any one time. For example, the use of a car by one driver prevents that car from being driven by anyone else concurrently. Intellectual property, by contrast, is non-rivalrous; due to its intangible nature, it can be used by any number of people at the same time (Arrow 1962). Consider, for example, a song: many people can sing that song simultaneously. *Exhaustibility* is a characteristic of physical property, whereby its quality or condition deteriorates over time. A car, for example, wears out through exposure to the elements or use. Intellectual property, however, does not wear out. A song sung one million times is no different from a song sung only once. *Excludability* is a characteristic of physical property, whereby the item may be kept in the sole possession of one person. A car, for example, may be locked, and it may be placed inside a locked garage, thereby making it very difficult for it to be driven by someone other than the owner. In contrast, intellectual property is non-excludable; because of its intangible nature, it is very difficult to lock up. A song, once published or performed, is "out there" for the world to sing, even if the song's owner does not consent to that.

Because of these characteristics, it is very difficult for someone to "possess" a piece of intellectual property as effectively as it is for someone to possess a piece of physical property. The provision of exclusive rights to intellectual property is the means by which the law makes possession of intellectual property possible. The nature of the laws that implement these intellectual property rights is explained next.

Characteristics of Intellectual Property Laws

In general terms, the laws that implement intellectual property rights have the following two key characteristics: the rights apply only in relation to a subset of all products of human intellectual activity; and the rights apply only in relation to a subset of all activities that might be undertaken in respect of those subject matters (Christie 2006, 27–28). The first characteristic means that not all products of the human intellect are made "possessable" by the law. The second characteristic means that, even where a product of the human intellect is made "possessable" by the law, the entitlements of possession are not unlimited.

Not every product of the human intellect is protected by intellectual property rights. Rather, it is only those intellectual products for which there is a specific legal regime of protection, and which satisfy the ingenuity threshold requirement of the regime, that obtain the benefit of the grant of exclusive rights. In essence, only those intellectual products that are either aesthetic or functional in nature are protected. Typical examples of *aesthetic* intellectual products protected by intellectual property law are works of art, literature and music, and the shapes of consumer goods. Copyright law is the regime by which protection is provided to works of art, literature and music, while design law protects the shapes of consumer goods. Typical examples of *functional* intellectual products protected by intellectual property law are scientific and technological inventions and trademarks (words or symbols that act as indicators of origin of product). Patent law and trademark law, respectively, are the regimes by which protection is provided to inventions and trademarks.

The various intellectual property rights regimes specify the subject matters to which they are applicable. For example, the exclusive rights of copyright may only be granted in respect of "works" or "neighbouring" subject matters, and the exclusive rights of a patent may be granted in respect of "inventions". The categories of subject matters protected by the different regimes are not mutually exclusive. For example, a computer program may be both a "work" (in particular, a literary work) for copyright purposes and an "invention" for the purposes of patent law (Christie 1994; Long 2009). In this case, the computer program will be protected under both copyright law and patent law, so long as it satisfies the ingenuity threshold requirements of each regime.

Each intellectual property regime has an ingenuity threshold that must be met before protection is afforded to any particular piece of intellectual property. Thus, only an "original" work may be protected by copyright. Likewise, only a "new" and "non-obvious" invention may be granted protection by a patent. The particular relevance of the concepts of creativity and innovation to ingenuity thresholds in intellectual property law is considered in detail in the section "Creativity and Innovation in Intellectual Property Law".

All of the intellectual property regimes provide the owner of a protected intellectual property subject matter with a number of exclusive rights. The regimes specify the uses of the subject matter that are the exclusive entitlement of the subject matter owner. For example, the exclusive rights of the owner of copyright in a work are the rights to reproduce, publish, perform, communicate and adapt the work. For a patented invention, the exclusive rights of the patentee are to "exploit" the invention, which includes making, using and selling the invention.

If a person exercises a specific exclusive right without the intellectual property owner's consent, that person is liable for infringement. Outside of the specific exclusive rights provided by each intellectual property regime, however, non-rights holders may use and interact with the subject matter as they like. Thus, it is not an infringement of copyright to *read* a literary work, because reading does not constitute an exercise of any of the copyright owner's exclusive rights (reproduction, publication, performance, communication, adaptation).

Rationales for Intellectual Property Rights

There are various rationales for the law's provision of exclusive rights to intellectual property. These rationales have one of two bases: morality and economics. Each of these bases emphasises a different set of values.

Morality-based rationales for intellectual property rights focus on the fact that creation and innovation involve an individual's labour, intellect and personality. These rationales seek to provide individual justice in dealings between members of society. Economics-based rationales, in contrast, focus on encouraging people to invest in creative and innovative activities. These rationales seek to reward creators and innovators for making available to the public the fruits of their investment. The various rationales have their own strengths as well as limitations, and there is a degree of overlap between each of them.

According to a *morality*-based rationale for intellectual property rights, individuals have a natural right of entitlement to the products of their intellectual activity and labour. Intellectual property rights prevent third parties from becoming unjustly enriched by "reaping what they have not sown". This is based on a corrective, distributive justice between the owner and the taker. A taker who "reaps" a reward that naturally belongs to a creator or an innovator should be punished.

Moral legal theorists are divided as to what it is, exactly, that entitles creators and innovators to protection over their creative and innovative products. Some believe that a moral justification for intellectual property rights is borne out of Locke's idea that creators and innovators have natural rights over the products of their labour (Hughes 1988; Child 1990). Others are influenced by Hegel's notion that works should be protected because they are an expression of a creator's or an innovator's personality (Radin 1982).

Both theoretical approaches have their weaknesses. For example, Locke's theory that a person acquires property rights where he or she exerts labour in relation to resources applies where those resources are either held in common by the state or are unowned. But, arguably, not all creators or innovators mix their labour with resources that are held in common or unowned (Himma 2006). In response to Hegel's theory that a person enjoys a right to the expression of their personality, it has been noted that certain types of intellectual property protected by the law—for example, inventions, integrated circuits and plant varieties—do not appear to reflect the personality of their producer (Hughes 1988).

The *economic* argument for intellectual property rights presupposes that without such rights, the production and dissemination of cultural, scientific and technological objects would not occur at an optimal level (Bently and Sherman 2004, 35). An economic rationale for intellectual property rights is that if creators and innovators are not given adequate legal protection over their works, many creations and innovations would not occur and society would suffer as a result. For example, an author would not invest the time and effort in writing a book to be enjoyed by millions of readers if, once the book was published, readers were free to copy it and other authors were free to plagiarise it. Likewise, a drug company would not spend many years and large sums of money developing a vaccine that saves millions of lives if, once the drug was released, its competitors were free to produce copies of it.

To provide the incentive for investment in creation and innovation, the law provides creators and innovators with exclusive rights to produce and sell the products of their creativity and innovation. This exclusivity, in turn, permits the intellectual property right's owner to commercially exploit the intellectual property for profit. Thus, the reward of exclusivity to the intellectual property resulting from creativity or innovation provides the incentive for its production.

The economic theories, too, have weaknesses. Some commentators question whether an incentive is necessary for the production and dissemination of at least some types of *aesthetic* intellectual works. If people enjoy the creative process and consistently express themselves in an artistic way, wouldn't they continue to do so irrespective of intellectual property protection? Also, some question whether there is enough done on the part of the creator to justify the reward of intellectual property protection. As has been noted, "copyright's threshold is set at a very low level and thus catches works which are created for their own sake, such as letters, holiday photographs, and amateur paintings" (Bently and Sherman 2004, 34). Furthermore, rewarding a creator or innovator with a monopoly to control the exploitation of his or her creation or innovation creates problems, as a monopolist tends to discourage use of a good by overpricing it. The dilemma is that without the legal monopoly provided by intellectual property rights not enough creations or innovations will be produced for society's benefit, but with the legal monopoly provided by intellectual property rights too little of those creations or innovations will be used by society (Cooter and Ulen 1988, 135).

CREATIVITY AND INNOVATION IN INTELLECTUAL PROPERTY LAW

Threshold Requirements of Ingenuity

As was noted earlier in the discussion of the nature of intellectual property rights, the laws that create intellectual property rights specify a threshold of ingenuity that must be satisfied for the subject matter to gain the benefit of the rights. Unless a particular piece of intellectual property satisfies the relevant threshold, the law does not provide any protection in respect of it. It is in relation to these ingenuity thresholds that the concepts of creativity and innovation have relevance in law.

Copyright law is primarily directed to the protection of aesthetic subject matters, such as works of art, literature and music. For such a subject matter to be protected under copyright, it must be "original". The concept of originality in copyright law varies around the world (Ricketson and Creswell 2002, 7.35). Some countries, notably continental European countries, understand originality to mean that aspect of the work that reflects the "personality" of the author. Other countries, notably common-law countries (United States, United Kingdom, Canada, Australia), adopt a somewhat lower standard. In these other countries, a reflection of authorial personality is not required; it will usually be sufficient that the subject matter resulted from some "independent effort" by the author—that is, the subject matter was not merely a copy of another work (Christie 2006, 30). Thus, the lowest common denominator of originality in copyright law is that the work be "not copied". Patent law, by contrast, is primarily directed to the protection of functional subject matters, such as products and processes. For a product or process to be protected by a patent, it must be both "new" (novel) and "inventive" (non-obvious). The characteristics of novelty and non-obviousness are judged against the "prior art". The "prior art" is all information publicly available anywhere in the world at the time of filing the application for a patent. It therefore includes all the information made public by the publication of documents and by the doing of acts in public (Christie 2006, 33).

The prior art invalidates an invention for lack of novelty if it discloses all features of the invention in clear and unmistakable terms. Prior art invalidates an invention for obviousness if it discloses features from which the invention differs only in obvious ways. Put the other way around, an invention is "new" if it does not exist in the prior art, and an invention is "inventive" if it is not obvious to a person skilled in the relevant prior art.

Scopes of Legal Protection

The scope of protection provided by an intellectual property regime is essentially a factor of the exclusive rights provided to an owner of a protected work and whether or not a causal connection is required for infringement. As noted earlier, both copyright law and patent law provide specific exclusive rights. There is a significant difference in the scopes of protection provided by copyright law, on the one hand, and by patent law, on the other hand. This significant difference is in respect of the *causal connection* requirement for infringement. This difference is explained in the following.

In copyright law, an exercise of the exclusive rights of the copyright owner will only be an infringement if there is a causal connection between the work protected by the exclusive rights of copyright and the subject matter in respect of which those rights have been exercised. Put more simply, there will only be an infringement if there is a causal connection between the protected work and the thing that allegedly infringes copyright in that work. In respect of the fundamental exclusive right of copyright, the reproduction right, the necessary causal connection for infringement is that the protected work has been *copied*. Thus, there is no infringement if the protected work is reproduced other than by copying—such as by chance or by derivation from a common source. That is to say, independent reproduction of the work is a defence to a claim of copyright infringement.

This characteristic of copyright law is illustrated by the following example. Person A and person B (whether contemporaneously or not) both make a sketch drawing of the Eiffel Tower. Both produce their drawing by looking at the Tower from the same location and, as a result, the two drawings are almost identical looking. Both drawings are protected by copyright, because they are both original—neither is copied from the other. Person C observes person A's drawing and makes a copy of it. Person C's drawing looks almost identical to both person A's drawing and person B's drawing. Although person C's drawing is almost identical to the drawings of both person A and person B, person C has infringed only person A's copyright. Person C has not infringed person B's copyright, because person C did not copy from person B's work.

It will be understood from this example that the essence of the exclusive right provided by copyright law is a right to *prevent copying*. Put simply, copyright provides a "copying" right—hence the name *copyright*. Under a copying right, *independent reproduction* of the protected subject matter is *not* prohibited.

The position under patent law is quite different. The exclusive rights of a patentee are infringed by one who makes the patented invention without the patentee's consent, whether or not there is a causal connection between the invention so made and the patented invention. Thus, infringement of a patent can occur *without copying*. That is to say, independent production of a patented invention is not a defence to infringement (Dufty and Lahore 2006).

This characteristic of patent law is illustrated by the following example. Person A invents a golf club with a double-sided head, such that it can be used by both left-handed and right-handed players. The invention is granted a patent because it is both new and inventive (not obvious) compared with the prior art (pre-existing golf clubs, which have only a single-sided head). Person B sees person A's invention and copies it. Unaware of both person A's invention and person B's copy of the invention, person C develops a similar golf club with a double-sided head. Both person B and person C infringe person A's patent, even though only person B copied person A's invention.

It will be appreciated from this example that the essence of the exclusive right provided by patent law is a right to be the *sole supplier* of the patented invention. Put simply, patent law provides a "monopoly" right. Under a monopoly right, any reproduction—*including independent reproduction*—of the protected subject matter is prohibited.

Role of Creativity and Innovation

Having considered and compared the threshold requirements in, and the scopes of protection provided by, copyright law and patent law, it is now possible to observe the relevance of the concepts of creativity and innovation in law. The relevance of these concepts is in the role they play as ingenuity threshold requirements in intellectual property law, particularly copyright and patent law.

The ingenuity threshold requirements in copyright law and in patent law operate to determine whether or not a particular subject matter may be granted protection under that regime. Given that the scopes of protection provided by copyright law and by patent law are fundamentally different, it follows that the ingenuity threshold requirements operate to determine which scope of protection is provided to an intellectual property subject matter that is one to which both regimes apply—such as a computer program. Where the subject is matter is merely "original", only the copying right protection under copyright is available. Where, however, the subject matter is "new" and "inventive", the monopoly right protection under patent law is also available.

The threshold requirement of "originality" in copyright law equates very closely to the concept of "creativity" in law. As noted previously, the ordinary, and thus the legal, meaning of "creativity" is the ability to bring something into being, whether or not the thing brought into being already exists. That is to say, creativity requires independent effort, but creativity does not require novelty or inventiveness. This is, in essence, the same as the ingenuity requirement of originality in copyright law, which requires that the work be independently created (i.e. "not copied").

In contrast, the ingenuity threshold requirement of "novelty" and "inventiveness" in patent law equates very closely to the concept of "innovation" in law. The ordinary, and thus the legal, meaning of "innovation" is the act of bringing into being something new. Whether or not something is new is relative; newness can only be determined by comparison. The necessary comparison is with those things that already exist. Thus, the concept of innovation is, in essence, the same as the ingenuity requirement of novelty and inventiveness in patent law, which requires that the invention be new and non-obvious compared with the prior art.

Drawing all this together, it will be seen that the concepts of "creativity" and "innovation" have a profound, and a significantly different, effect in law. The concept of "creativity" is one that justifies the grant of *copying* rights in respect of intellectual property subject matters. In contrast, the concept of "innovation" is one that justifies the grant of *monopoly* rights in respect of such subject matters. This consequence is captured in Table 7.1.

Table 7.1 Creativity and Innovation in IP Law

	Creativity	Innovation
Legal meaning is:	the ability to bring something into being	the action of bringing something new into being
Equates to threshold requirement in:	copyright law	patent law
Results in IP protection scope of:	copying rights	monopoly rights

POLICY IMPLICATIONS

Taking into account all of the preceding, the policy outcome of the law's perspective on creativity and innovation is clear: different types—and, more importantly, different strengths—of legal protection are provided for the exercise of different types of intellectual ingenuity. Where mere creativity (the lower level of ingenuity) has been exercised, a lower level of legal protection is provided (typically through the copying rights of copyright). Where, however, innovation (the higher level of ingenuity) has occurred, the law provides a stronger level of protection (typically through the monopoly rights of a patent).

There is a very important consequence of the law providing different strengths of protection depending on the level of ingenuity exercised. This consequence concerns the effect of the differing strengths of protection on the activities of subsequent, or "follow-on", creators and innovators—that is, on those who seek to build on pre-existing creations and innovations. By providing different strengths of protection, the law imposes different degrees of constraint on follow-on creators and innovators depending on the level of ingenuity exercised by earlier creators and innovators. In particular, the constraints imposed on follow-on creators and innovators are greater where the earlier producers of intellectual property have exercised greater levels of ingenuity.

The reason this consequence arises is because a monopoly right can be exercised against a wider range of persons than can a copying right. In particular, a monopoly right can be exercised against those who reproduce the protected subject matter independently as well as by copying it, whereas a copying right can only be exercised against those who reproduce the protected subject matter by copying it. Thus, independent reproduction will not save a follow-on creator or innovator from infringement of the legal protection provided to an earlier producer of intellectual property who has been innovative, not just creative—that is, who has not just brought something into existence, but who has brought into existence something that is new (not previously existing).

The question that arises in respect of this consequence is whether it is justifiable to constrain follow-on creators and innovators more in respect of past innovations than in respect of past creations. To answer that question one must consider the impact of the different degree of constraints from two perspectives: that of the follow-on creator/innovator and that of the earlier creator/innovator. From the perspective of the follow-on creator/ innovator two factors may be noted. First, the *duration* of the higher degree of constraint is significantly shorter than is the duration of the lower degree of constraint. The duration of the exclusive rights of a patent is twenty years from the date of filing of the patent application. In contrast, the duration of the exclusive rights of copyright is, in general, seventy years from the death of the creator of the copyright work. Secondly, the *likelihood* of the higher degree of constraint being imposed on the follow-on creator/ innovator is significantly lower than is the likelihood of the lower degree of constraint being imposed. This is simply a function of the ingenuity threshold. In general terms, while the exercise of ingenuity is likely to result in something being brought into existence, it is much less likely that the thing being brought into existence is new. Thus, the exercise of ingenuity is much less likely to result in the production of intellectual property to which the higher degree of protection applies. Accordingly, it is much less likely that follow-on creators/innovators will have imposed on them the higher degree of constraint than the lower degree of constraint. From the perspective of the original creator/innovator it may be noted that the law provides a much more valuable *reward*—whether seen as an entitlement under a morality-based rationale for intellectual property protection or as an incentive under an economics-based rationale for intellectual property protection—for the exercise of the higher level of ingenuity. The more valuable entitlement, and the stronger incentive, of a monopoly right is available only where the intellectual ingenuity involved goes beyond mere "creativity" and amounts to "innovation".

These three factors, in combination, would seem to justify the policy outcome of differing strengths of protection, with consequential differing degrees of constraint on follow-on creators/innovators, for the exercise of different levels of ingenuity. The negative impact of the higher degree of constraint on follow-on creators/innovators is balanced by the fact that the higher degree of constraint is less likely to arise than is the lower degree of constraint; and by the fact that when the higher degree of constraint does arise, it lasts for a shorter period than does the lesser degree of constraint. Furthermore, there is a valid justification for providing the higher degree of constraint in those situations where it does arise. The justification is to reward innovators for exercising the higher degree of ingenuity. This justification is valid because, while it is beneficial to society that things be created, it is clearly more beneficial to society that at least some of the things being created are *new* to society. Without the creation of new things, society would not advance culturally or economically. Thus, it is justifiable to provide a more valuable entitlement, and a stronger incentive, for innovation compared with mere creativity.

CONCLUSION

The legal perspective on creativity and innovation is simple but profound. Neither creativity nor innovation has any particular relevance in law, except in respect to the role those concepts perform in intellectual property law. Intellectual property law provides protection to aesthetic and functional emanations of the human intellect, by granting exclusive rights over them to those responsible for producing them. These rights both recognise the moral entitlements of creators and innovators over their intellectual property and provide an incentive to creators and innovators to produce intellectual property.

In intellectual property law, creativity and innovation are alternative thresholds of intellectual ingenuity that must be satisfied for intellectual property subject matter to gain protection. Where an intellectual property subject matter results from the exercise of creativity, the law provides the protection of a copying right. Where, however, the subject matter results from a higher degree of intellectual ingenuity, such that it amounts to an innovation, the law provides the stronger protection of monopoly rights. This outcome reflects a policy that the production of *new* things is more valuable to society than the mere production of things.

BIBLIOGRAPHY

Arrow, Kenneth J. 1962. "Economic welfare and the allocation of resources for invention". In *The rate and direction of economic activities: Economic and social factors*, ed. R. Nelson, 609. New York: Princeton University Press.

Bently, Lionel, and Brad Sherman. 2004. *Intellectual property law second edition.* Oxford: Oxford University Press.

Canadian Oxford Dictionary Second Edition. 2004. Toronto: Oxford University Press.

Child, James W. 1990. The moral foundations of intangible property. *The Monist* 73: 578–60.

Christie, Andrew. 1994. Designing appropriate protection for computer programs. *European Intellectual Property Review* 16: 486–93.

———. 2006. "Intellectual property and intangible assets: A legal perspective". In *The management of intellectual property*, ed. Derek Bosworth and Elizabeth Webster, 23–39. Cheltenham, UK, and Northampton, MA: Edward Elgar.

Cooter, Robert, and Thomas Ulen. 1988. *Law and economics.* Glenview, IL: Scott, Foresman and Company.

Dent, Chris. 2006. Patent policy in early modern England: Jobs, trade and regulation. *Legal History* 10: 71–95.

Dufty, Ann, and James Lahore. 2006. *Patents, trade marks and related rights.* Sydney: LexisNexis Butterworths.

Hay, David, ed. 2007. *Words and phrases legally defined*. London: LexisNexis Butterworths.

Himma, Kenneth Einar. 2006. The justification of intellectual property: Contemporary philosophical disputes. *Journal of the American Society for Information Science and Technology* 59: 1143–61.

Hughes, Justin. 1988. The philosophy of intellectual property. *Georgetown Law Journal* 77: 287–366.

Long, Doris Estelle. 2009. At the boundaries of access when worlds collide: The uneasy convergence of creativity and innovation. *John Marshall Journal of Computer & Information Law* 25: 653–71.

Macquarie Dictionary Federation Edition. 2001. Sydney: Macquarie Library.

Martin, Elizabeth, and Jonathan Law, eds. 2006. *A dictionary of law*. New York: Oxford University Press.

Merriam-Webster Dictionary Online. http://www.merriam-webster.com/.

Oxford English Dictionary Second Edition. 1989. Oxford: Oxford University Press.

Phillips, Jeremy, and Alison Firth. 1995. *Introduction to intellectual property law third edition*. London: Butterworths.

Radin, Mary Jane. 1982. Property and personhood. *Stanford Law Review* 34: 95–1015.

Ricketson, Sam, and Christopher Creswell. 2002. *The law of intellectual property: Copyright, design & confidential information*. Sydney: Lawbook Company.

Sawyer, R. Keith. 2008. The shape of things to come: Creativity, innovation and obviousness. *Lewis & Clark Law Review* 12: 461–86.

Sherman, Brad, and Lionel Bently. 1999. *The making of modern intellectual property law: The British experience: 1760–1911*. Cambridge: Cambridge University Press.

8   Promoting Creativity and Innovation through Law

Brian Fitzgerald

INTRODUCTION

Innovating entails the use of the creative endeavour to find and then apply new ways of doing things (Cutler 2008b, 15). Creativity involves the origination of new ideas yet not necessarily their practical application. The distinction may be hard to maintain at points. For instance, if I think of a new way to provide a Web service and post those thoughts to a blog or my Facebook page have I simply been creative or can it be said that I have also been innovative? The important aspect to understand is that innovating relies on creativity to prosper. If we cannot promote creativity we can never hope to provide the opportunity for innovation (Schumpeter 1943, chap. VII; Cutler 2008b, 15–23; 2008a; Howkins 2001).

The law has a fundamental role to play. In the twenty-first century we have seen the emergence of vast online social and cultural networks in which millions of people exchange ideas everyday. From Wikipedia to YouTube to Facebook we have seen how the peer producer, user generator and social networker are changing the face of knowledge and creativity (Benkler 2006; Jenkins 2006). The network is open, distributed, serendipitous, transnational and much more. At the core of this revolution is the notion of maximising the broadest possible exchange of ideas and ensuing freedom to operate with those ideas (Shapiro and Varian 1999; Chesborough 2003).

The legal system needs to be responsive to these changes if it is to be a facilitator rather than an obstacle for innovation. It needs to provide a context or framework for this dynamic exchange of ideas to occur. The legal system has long cherished the ideal of freedom of speech as an aspect of social and political activity; however, today freedom of thought, the exchange of ideas and the ability to implement those ideas in an experimental way are the core ingredients of innovation (Cutler 2008b; Metcalfe 2008). The more we can exchange ideas and experiment without fear of retribution (be creative) the more we can innovate—that is the realisation of the twenty-first century.

The legal system needs to promote this dynamic by ensuring the "freedom to innovate" (Jefferson 2007). To do this the law must (at very least) ensure:

1. freedom of expression and communication
2. "end to end" networks (like the Internet)
3. an information environment which allows experimentation with and implementation of disruptive communication technologies
4. a greater freedom to reuse and transform existing knowledge

In this chapter I explore ways in which the law might better promote innovation by guaranteeing these facilitators of creative endeavour and ultimately innovation. In doing so I posit four core principles that underpin the discussion and in turn provide a basis for assessing the effectiveness of the law.

FREEDOM TO COMMUNICATE: FREEDOM TO INNOVATE

Principle 1—Make Sure We Can Communicate Ideas to and amongst Each Other to the Greatest Possible Extent under Law

The online social networks of the twenty-first century evidence an exchange of ideas of great magnitude and geographic reach. Participation in such networks requires at the very least a capacity to access the network (economically and physically) and an ability to communicate and participate once on the network (digital literacy).

Traditional notions of freedom of expression are important in this environment. In Australia the Constitution only guarantees a right to freedom of political communication (*Lange v. Australian Broadcasting Corporation* [1997] HCA 25; Fitzgerald 1993). Any other legal protection of freedom of communication must be found in the principles of general law (statutory or common law) or the unregulated parts of social activity. Other countries protect the right to freedom of expression to varying degrees.[1]

The study of constitutional law in Australia since federation has focused on the workings of the state and particularly the *separation of powers* between the different organs of the state (legislature, executive and judicial) and the *division of powers* between the Commonwealth and the states as part of the Australia federal system. Procedural issues concerning the operation of the key actors of the state—for example, parliament and specifically the power of the Australian upper house or Senate to block "supply" or appropriation/money bills or the governor general's right to dismiss a prime minister—have also taken centre stage. All are of the utmost importance but an inability by the discipline to more fully examine the politics of power—and how the law should respond—beyond the organs of the state has left the study of constitutionalism in Australia a fairly narrow exercise.

For a period in the 1990s the High Court of Australia attempted to widen our legal understanding of these issues by implying rights into the Constitution. However, this short burst of enthusiasm for implied rights has been somewhat tempered by the current High Court, which has been cautious about such an approach. On being appointed attorney general for the Commonwealth of Australia The Hon. Robert McClelland reiterated the Rudd government's commitment to examining whether a bill of rights should be introduced in Australia. He explained that any such proposal would not entail the judiciary having the power to invalidate laws passed by parliament but rather they would be able to issue opinions on validity that would form part of the pressure on parliament to redress the breach (Pearlman 2007). This would most likely be a statutory as opposed to constitutionally entrenched bill of rights similar to those existing in Victoria, the ACT and New Zealand.

Michel Foucault taught us that power relations occur everywhere in our daily lives and that they are not simply the exclusive domain of citizen and state relations (Fitzgerald 2007; also see The Foucauldian Library at http://web.archive.org/web/20080602065658; http:/www.thefoucauldian.co.uk/library.htm). The Critical Legal Scholars (CLS) (Kelman 1987), a latter-day version of the American Realists (Fisher, Horwitz and Reed 1993), built on the Foucauldian notion of power, arguing that the law governing private relations is as important and politically dynamic as the law governing citizen and state relations. This unique insight was largely ignored by constitutional law scholars. The sentiment was either one of straight-out rejection of the philosophical approach or that if we went down that road everything would be a question of constitutional law (cf. Fitzgerald 1994). The taxonomy was set in stone and this was not constitutional law. To this end the boundaries of the common law and statutory obligations concerning matters such as contract, tort, intellectual property and competition remained firmly entrenched as questions of private law not constitutional law.

This approach has always seemed problematic and short-sighted. The rise of a networked information society wherein private companies build regulatory regimes through contract and intellectual property and enforce these through self-executing technological implementation measures has only served to heighten these concerns (Fitzgerald et al. 2007). Our ability to communicate is now intimately connected with software, computers, the Internet and the private companies that control their reproduction, population and dissemination. Elsewhere I have written about what I call "software as discourse" (Fitzgerald 2000). By this I mean that software is an instrument which allows us to understand things and, in the sense that Martin Heidegger used the word, software is a form of "discourse" (Heidegger 1962).[2] How we protect rights in and around software development, dissemination and usage is critical to our ability to construct knowledge and meaning.

In this context free speech has grown from a basic right in relation to government to raise my political concerns on a soap-box in the marketplace to one that now serves to promote my ability to exchange ideas in the vast online social networks of the twenty-first century. Government is still a central concern but just as important is the reach of the private corporation in shaping what I can do. As private corporations are mandated to maximise their self-interest the great concern is as to how public values—constitutional values—are inserted in this framework (Lessig 1999; Berman 2000).

As the greatly respected English jurist A. V. Dicey, a professor of law at Oxford University, explained over 125 years ago, our rights are protected by both the general law and what is left unregulated. In relation to free speech the boundaries of the law relating to things such as defamation and sedition have a role in shaping our ability to engage in free speech. In the digital world the ability to use different types of software products (competition law[3]) and to be able to remix and reuse content as a part of cultural existence (copyright law) are key issues.

What should the law do and where should it take us? To start with we need to try and state what the core principles that underpin our constitutional compact are—what I term principles of constitutionalism. Values such as free speech, equality, respect, fairness, reward, fair use and so on. We need to invigorate and infuse the development and interpretation of the general law with the core principles that underpin our society to, as A. V. Dicey explained, properly articulate our basic rights and freedoms (Dicey 1975). Such an action is made even more urgent by the rise of the networked information society and the large role private actors play in its construction and governance (cf. *Bragg v. Linden Research Inc.* 487 F. Supp. 2d 593 [E.D. Pa. 2007]).

Layer on top of this the notion that the exchange (including the transmission) of ideas is a key driver of innovation, that innovation improves productivity and that Australia is desperately trying to improve its productivity. How do we draw this link between innovation and constitutionalism; between the need to provide development and interpretation of the law in a way that as far as possible promotes the exchange of ideas in the name of innovation? The term "freedom to innovate" is one that Dr. Richard Jefferson has used in recent times (Jefferson 2006). It sums up the joining of constitutional values with innovation policy in the twenty-first century. What we are aiming to do is to provide a freedom to exchange ideas and to experiment in order to promote innovation—the freedom to innovate.

One key point that emerges from this analysis of the legal context of this first principle is that we need to work harder at understanding how the development and interpretation of the general law acts as part of our constitutionalism and in turn how critical that general law is to promoting creativity and innovation.

END TO END NETWORKS

Principle 2—Make Sure the Network Is Kept as Open as Possible to Allow the Most Dynamic Exchange of Ideas

The Internet is said to be based on an end to end (e2e) design system so that intelligence and customisation occurs at the end-points rather than in the middle (Lemley and Lessig 2000; Fitzgerald and Fitzgerald 2002, chap.

1–3; Fitzgerald 2007). This is evidenced by the fact that the Internet is made up of a series or stack of protocols in which each layer is independent of the next with the higher level (closest to the user) being more customised. This idea has an analogy in pre-existing analogue networks like that of the telephone. In making a phone call the message flows unfiltered through the wires or pipes to the other end where devices are attached to make it understandable. Interference through monitoring or editing is not seen as a part of the design system.

The free-flowing and distributed nature of the Internet is what makes it such a unique and dynamic communication network. Any tampering with the design of the system has the potential to impact on the flow and exchange of ideas. We all understand that no right is absolute (*Schenck v. United States*, 249 US 47 at 52 per Justice Holmes), and that the law balances freedoms against social order (e.g. Schedule 5 of the *Broadcasting Services Act 1992* [Cth]). But where the only reason for interfering with the design system is commercial gain, the question becomes: to what extent should the fundamental freedom of the network be protected under law?

In recent years the issue of "Net(work) Neutrality" has gained significant attention.[4] This topic covers many aspects but one simple example of this argument is found in the situation where Internet Service Providers (ISPs—companies that provide Internet access) employ routers (computer technology that forwards or sends Internet messages/data packets through the network) to differentiate between and, where a commercial incentive exists, speed up or slow down data streams. For example, ISPs in recent years have explained that peer to peer (P2P) traffic (music, film, games, phone calls) is taking up a large amount of available bandwidth and costing them a lot of money with little commercial return. One strategy is to structure Internet behaviour by slowing down the P2P traffic and providing the more commercially attractive data streams with a faster service. This approach conflicts with the basic principle of e2e design: that the data should flow unimpeded through the network and be managed at the ends.

As suggested earlier the purest exchange of ideas in social networks of the twenty-first century is an important part of the knowledge-construction process (Hartley 2007). In the words of Vint Cerf, one of the founders of the Internet, "a lightweight but enforceable neutrality rule is needed to ensure that the Internet continues to thrive". He explains:

> The remarkable social impact and economic success of the Internet is in many ways directly attributable to the architectural characteristics that were part of its design. The Internet was designed with no gatekeepers over new content or services. The Internet is based on a layered, end-to-end model that allows people at each level of the network to innovate free of any central control. By placing intelligence at the edges rather than control in the middle of the network, the Internet has created a platform for innovation. This has led to an explosion

of offerings—from VOIP to 802.11x Wi-Fi to blogging—that might never have evolved had central control of the network been required by design. (Cerf 2005)

In tackling the question of Net Neutrality some have suggested that we should pass laws to reinforce it[5] while others have cautioned against such an action. Existing competition or anti-trust laws have a role to play where an entity misuses significant market power, yet Net Neutrality is not always a simple question of an abuse of a dominant position in the marketplace.

More generally, and in line with the argument made in the previous section, in approaching the development and interpretation of the law we should be mindful that creativity, innovation and ultimately productivity can prosper in an environment where the dynamic exchange of ideas is facilitated by a free-flowing network (Obama and Biden 2008)

AN INFORMATION ENVIRONMENT THAT ALLOWS EXPERIMENTATION WITH AND IMPLEMENTATION OF DISRUPTIVE COMMUNICATION TECHNOLOGIES

Principle 3—Provide Opportunity for Experimentation with and Implementation of New and Disruptive Communication Technologies

One of the most critical issues for an innovation system concerns change and renewal (creativity) within the system. If everything since time began had remained the same then we would not be living—for good or bad—the way we are today. Change provides new opportunities, new markets and new solutions but it is often resisted if not stifled at its inception. The challenge for an innovation system is to ensure that it promotes and sensibly accommodates new ways of doing things even if they are disruptive.

As Joseph Schumpeter, the doyen of innovation theory, explains: "But in capitalist reality . . . it is not the kind of competition which counts but the competition from the new commodity, the new technology, the new source of supply, the new type of organisation . . . competition which commands a decisive cost or quality advantage and which strikes not at the margins of the profits and the outputs of the existing firms but at their foundations and their very lives" (1943, 84).

Over the last ten years we have seen this fundamental tenet of the innovation process tested in the development of the Internet, especially in the area of copyright law. The traditional copyright industries based around music and film have sought to shut down through legal action new forms of distributing or communicating digital entertainment products across the Internet. Starting with *A & M Records Inc. v. Napster Inc.* (239 F. 3d 1004 [9th Cir. 2000]; [*Napster*]) through to *Universal Music Australia Pty Ltd.*

v. Sharman License Holdings Ltd. ([2005] 65 IPR 289; [2005] FCA 1242; [*Kazaa*]) and *MGM Studios Inc. v. Grokster Ltd.* (545 US 913 [2005]; [*Grokster*]) we have seen the recording industry successfully pursue intermediaries that developed and/or distributed P2P file-sharing technology or software through the courts (Fisher 2004).[6] The defendants in these cases for the large part were not knowingly or intentionally reproducing or communicating unauthorised copies of songs but rather providing the facilities and services for others to do so.

Under copyright law you can infringe copyright by actually "doing" the infringing act (primary liability) or by authorising another person to do an infringing act (authorisation or secondary liability). In the United States, secondary liability is spoken of in terms of contributory, inducement or vicarious liability. To authorise means to "countenance, sanction or approve" (*University of New South Wales v. Moorhouse and Angus & Robertson* [1974] 133 CLR 1) and is intimately linked with the notion of control. Sections 36 (1A) and 101 (1A) of the *Copyright Act 1968* provide non-exhaustive statutory criteria for determining whether authorisation has occurred. Those sections explain that "in determining . . . whether or not a person has authorised the doing in Australia of any act comprised in the copyright . . . without the license of the owner of the copyright, the matters that must be taken into account include . . . (a) the extent (if any) of the person's power to prevent the doing of the act concerned; (b) the nature of any relationship existing between the person and the person who did the act concerned; (c) whether the person took any reasonable steps to prevent or avoid the doing of the act, including whether the person complied with any relevant industry codes of practice" (*Copyright Act 1968*, Section 36).

In a user-generated, distributed Web 2.0 world the notion of intermediaries authorising end-user infringement is a little far-fetched (*CBS Songs Ltd. v. Amstrad Consumer Electronics PLC* [1988] 1 AC 1013). The whole idea of the model is to allow the end-user to drive the system. The more we seek to control the network in the name of property rights the more we limit its potential.

If we conceptualise copyright law as not only being about making copyright owners wealthy in the name of the public good but also about enhancing wider economic, democratic and cultural goals then an authorisation doctrine that does not accommodate disruptive and innovative technologies—to some degree—is failing the innovation system more broadly.

It would be radical to argue that there should be complete immunity from liability, but at the very least we need to have the debate as part of legal argument as to where the boundaries should be drawn.

Since the question as to whether authorisation has occurred in any given case is one of fact and the statutory criteria in Sections 36 (1A) and 101 (1A) are not exhaustive, it seems inevitable that in future cases the assessment of authorisation must take into account the need for, and ubiquity and value of, user-driven distributed information sharing technologies in social discourse, creative innovation and the knowledge economy. The notion of authorisation is an important regulatory point within any copyright regime. What falls within its boundaries in essence extends the reach of copyright ownership, and what falls outside of it allows a greater number of unrestricted communicative activities. In this way the definition of authorisation, like that of other copyright fundamentals such as idea/ expression, substantial part and fair dealing, is critical to the free flow of ideas, innovation and democracy and determines what activity falls within or outside the public domain or copyright (Fitzgerald et al. 2007, 213–14). The frightening aspect of the P2P litigation has been the ease with which the recording industry has gotten its way and the inability of the judges— with the exception of Justice Stephen Breyer in the US Supreme Court in *Grokster* (545 US 913 at 949–66 [2005]—to see the big picture. More so, some have suggested that the litigation strategy of the recording industry has threatened the existence and adoption of new business models and new technologies (Lessig 2002). This is a matter of real concern as what we are witnessing is truly unique and a monumental change in social interaction. Never before have we seen communication on this scale and with such informality; millions of people forming an instantaneous and worldwide network for sharing knowledge. This is the very engine of creativity that an innovation system would crave, yet established industries are quick to try and limit its significance.

We have to be vigilant in our goal to ensure this type of freedom to innovate. The law of copyright, as suggested earlier, needs to be interpreted in a way that accommodates such inspirational change. What is more, we could look at how the copyright law might go further and impose obligations on copyright owners for the negative externalities they produce. A clear analogy exists in real property law. One hundred years ago real property or landowners had the right to use their property as they wished. The rise of environmental law over the last sixty years has seen this sovereign right of the landowner subjected to a series of obligations to ensure land use does not pollute the existing environment to the detriment of the general public.

Large entertainment companies holding intellectual property (particularly copyright) have steadfastly refused to promote new modes of exchange (for example, P2P technologies) and instead have pursued a line of suing intermediaries for distributing new technologies. They have asserted their sovereign right to exercise their property rights in any way they wish regardless of negative externalities. However, the information environment like the natural environment is an ecosystem. As the argument would go, by trying to stifle or kill off the emergence of new communication structures established industries have polluted the stream of the information ecosystem (cf. Boyle 1997).

If the law is to sponsor creativity in the vast networks of the Internet we need to see a much more sensible approach to secondary liability because if the boundaries are drawn too widely the potential for innovating will be dramatically restricted. There is no better example of this than in the multi-billion dollar lawsuit Viacom (representing the interests of Hollywood) has taken against YouTube (owned by Google) for alleged copyright infringement.[7] Google Inc., who has suggested that this litigation will determine the future of the Internet,[8] is the leader of a new breed of what I call "access corporations" that profit from greater access to knowledge—the more access there is the more money they make. YouTube is a classic example of this, being built around freely accessible, short, user-generated videos that are situated in a giant advertising scheme that earns Google enormous amounts of revenue. Should Google through YouTube be able to provide these services regardless of the fact that the user generators are appropriating material from Hollywood? This is a difficult dilemma for the law to resolve. However, if, as practice shows, we are moving from a control mode of distribution to an access model, how much value can the law add by constantly denying this shift in the way we live and act? The remedy for the traditional copyright industries should be found in the marketplace, not the artificial application of the law. A decision in favour of Google in this case would send the clear message to the traditional copyright industries to engage with the new distribution environment. To this end a more purpose-driven (yet limited) notion of secondary liability for copyright infringement is critical to the emergence of new technologies in the digital environment.

Some support for this view can be derived from the reasoning of the High Court in its recent decision in *Northern Territory v. Collins* ([2008] HCA 49), which considered the analogous issue of contributory or secondary infringement in relation to patents. Section 117 (1) of the *Patents Act 1990* provides that "if the use of a product by a person would infringe a patent, the supply of that product by one person to another is an infringement of the patent by the supplier unless the supplier is the patentee or licensee of the patent." Section 117 (2) explains that this will not cover (without more) the supply of a "staple article of commerce" or, in other words, an everyday item of commerce. In the words of Justice Crennan (at 144), staple articles of commerce are "products with significant non-infringing uses, or as it has been put in relation to the American provisions, products with 'lawful as well as unlawful uses'". This means that if the product supplied is an everyday item of commerce and capable of being used in a non-infringing way, then no liability should arise, as to hold otherwise would give the patent holder an enormous power to control everyday commerce and life.

In *Collins* it was alleged that pursuant to Section 117 (1) the Northern Territory government had by supplying cypress pine to a third party infringed a patent held by Collins relating to the extraction of essential oils from cypress pine. The High Court held that the supply of the cypress pine was the supply of a staple article of commerce and therefore liability did not arise—see Section 117 (2) (b). In so holding the High Court allowed room for the supply of everyday goods used for social and economic endeavours. It would be interesting to see if this same logic would remain if the everyday product being supplied was file-sharing software or content hosting and the action was for copyright rather than patent infringement.

THE POWER OF REUSE: FREEDOM TO REUSE AND TRANSFORM EXISTING KNOWLEDGE UNDER IP LAW

Principle 4—Provide Broader Opportunity for Existing Culture and Knowledge to Be Reused

The ability and capacity to creatively reuse information, knowledge and culture is a key component of innovation strategy (Cutler 2008b; Mayo and Steinberg 2007; Fitzgerald 2008a; Wunsch-Vincent and Vickery 2007).

Copyright

However, copyright law through its requirement to gain permission before using copyright material in a way that comes within the exclusive rights or control of the copyright owner creates an impediment for the dynamic and serendipitous reuse of copyright material that Web 2.0 allows (O'Reilly 2005). This can be addressed in many ways but two seem obvious. Firstly, we need to adopt more flexible yet economically viable reuse rights. Secondly, we need to look at how we can free up publicly funded copyright material for reuse. Much of the publicly funded copyright that exists in Australia and elsewhere (for example, government or crown copyright) could be made more accessible through more dynamic Web-based licensing models (like Creative Commons) that provide permission in advance.[9]

In order to better understand these suggestions in relation to copyright law and practice let us consider the following examples of transformative and crown use and then look at how government is responding to the challenge of making their data more accessible and usable.

Transformative Use[10]

Green Day is a popular punk rock band from San Francisco. Their 2004 CD titled *American Idiot* made them one of the most popular bands in the world. Dean Gray is a group from Perth in Western Australia. In early December 2005 they uploaded to the Internet a remix version of Green Day's CD *American Idiot* styled *American Edit*. Within days they received a cease and desist letter from Warner Brothers and Green Day. Ironically, Green Day portray themselves as a pop/punk band that questions the current order of things—singing lyrics such as "subliminal mind f*** America".[11] Dean Gray's remix is interesting and to many it will be refreshingly creative. Dean Gray is like many of a new generation of amateur creators. They can sit at home in the bedroom and produce the most wonderful things. Most often they do not want money. Merely they wish to share the finished product with the world. How do we harness the creative energy of the Dean Grays across the distributed landscape of the Internet? A critical issue in economic and social terms becomes whether there is anything wrong with Dean Gray's creative expression. Have they robbed the sound recording corporation of an opportunity in the derivative market? Have they "ripped off" the reputation and notoriety of Green Day? Have they made money from this endeavour? Are highly transformative non-substitutable derivatives such a bad thing for our culture? One solution is to more sensibly and adequately define the rights to engage in transformative (re)use of copyright material.[12]

Crown Use

Another scenario is that raised by the recent High Court decision in *CAL v. NSW* ([2008] HCA 35), which suggests that copyright material deposited in public registries (in that case survey plans of house blocks deposited in the land title registry) cannot be reused by government or the community without permission or remuneration (Fitzgerald and Atkinson 2008). This decision has the potential to provide another layer of taxation on access to information supplied to government registers or under reporting schemes (e.g. *National Greenhouse and Energy Reporting Act 2007*; see http://www.climatechange.gov.au/government/initatives/national-greenhouse-energy-reporting.aspx) and to thereby create a disincentive to reuse this existing information. Ultimately the question will need to be answered as to whether such an impost places too high a price on innovation when in many cases the creator of the copyright material will have already received an enormous benefit or privilege by being included in or complying with the public register or reporting scheme.

Open Access to Public-Sector Information (PSI)

As mentioned earlier, yet another approach is to look at how we can improve the licensing practices (the permission-giving process) in the public sector to make more publicly funded material available for reuse. Australian governments at all levels hold enormous amounts of material (data in the broader sense of the word) covered by copyright. Increasingly, citizens desire better access to this huge, publicly funded, yet largely inactive archive to source creativity and innovation in solving key intergenerational issues such as climate change, water management and child health, building new services and industries and reinvigorating Australian culture through, for example, a Creative Archive.[13] As Lauriault and McGuire (2008) explain:

> There is a global movement to liberate government-"owned" data sets, such as census data, environmental data, and data generated by government-funded research projects. This open data movement aims to

make these datasets available, at no cost, to citizens, citizen groups, non-governmental-organizations (NGOs) and businesses ... Currently, access to government data is hampered by four main factors: i) the high cost of available data sets; ii) arbitrary decisions about availability of data sets to the public; iii) restrictive licenses; and iv) inaccessible data formats ... Liberating data will spur grassroots research on important social, economic, political and technical areas, currently hampered by lack of access to and high cost of civic data. Further, we want to link the debate about data to questions of government transparency and accountability, which pivot on access to accurate, reliable, and timely data.

The Australian Minister for Finance, Lindsay Tanner, has highlighted this changing role of citizens in the context of government. He has suggested that "the rise of internet-enabled peer production as a social force necessitates a rethink about how policy and politics is done in Australia", and that "there are significant opportunities for government to use peer production to consult, develop policy and make closer connections with the citizens." Minister Tanner explains that "as a huge creator and manager of information with an obligation to be open and transparent, we [government] have little choice" (Dearne 2008).

The Government Information Licensing Framework (GILF)[14] based in Queensland has started to deal with the legal issues facing "Government 2.0" and is now implementing the use of Creative Commons (open content licences) in relation to government data. As the permission-based nature of copyright fits awkwardly with the "real-time, fluid and serendipitous nature of modern information networks, Creative Commons licences have arisen to provide a Web-enabled system for providing permission in advance. Through the use of labelling they inform the downstream user as to what rights the copyright owner is giving them in advance. The permissions are based on one or a combination of the four primary protocols or conditions:

1. You can reuse my material so long as you attribute me—BY.
2. You can use my material for non-commercial purposes only—NC.
3. You must share improvements back to me—share alike SA.
4. You cannot alter my material in any way—no derivatives ND.

The licences are voluntary, meaning they will only be used where the copyright owner wants to use them, yet they have become an international standard for sharing or licensing copyright material in a lawful and open manner.

It is expected that Creative Commons–style open content licences will be adopted across the government sector to provide greater access to publicly funded resources as has happened in the research sector (Fitzgerald et al. 2006; Cutler 2008b). As Cutler's *Venturous Australia* report highlights, there is a desperate need for clear and strategic national information policy to provide coordination and leadership in this area (Cutler 2008b, Recommendation 7.7). The Ministry of Justice in the government of Catalonia in Spain has explained their adoption of CC in the following way:

> Nowadays the Internet is about sharing, co-producing, transforming and personalizing to create new products and services. To create, it is necessary to be able to make use of knowledge that already exists, without limits, and to share it afterwards. This is the philosophy of innovation that is now all-pervasive thanks to the democratization of technology. Creative Commons (CC) licenses are legal texts that allow authors to hand over some rights of their work for the uses they deem appropriate. So, these licences are an alternative for managing the author's copyright in a more flexible way. As a public Administration, the Ministry of Justice has decided to use CC licenses with the idea of turning over the knowledge created by the organization to the public so that it can be re-used. In this regard, CC licenses have been essential for this opening-up of knowledge. (Graells et al. 2008)

Cutler's *Venturous Australia* report has likewise recommended that "material released for public information by Australian governments should be released under a creative commons licence" (Recommendation 7.8).

Reforms

Fundamental reforms that would go a long way to making Australia a leader in terms of allowing reuse of copyright material—in order to sponsor creativity and innovation—would be the introduction of clear:

- rights to reuse copyright material for non-commercial purposes in circumstances where there is no financial detriment to the copyright owner
- rights to engage in transformative and fair use
- rights to reuse crown copyright for non-commercial purposes
- rights to engage in format shifting, e.g. changing material from analogue to digital format or changing material from one digital format (CD) to another (MP3) (Fitzgerald 2008b)

Patent

On another front the exponential growth of patent applications has raised concerns about the efficiency of the patent system and its very reason for existence. Patent offices around the world are being swamped by millions of patent applications each year with over four hundred thousand applications expected in the United States this year. The ability of examiners worldwide to cope with this growth and to adequately assess the newness, novelty and inventiveness of new technologies has been called into question.

An immediate response is to seek amendment of the patent law to make it more difficult to obtain a patent. This may happen in due course; however, I would suggest that there are two strategies that can be employed immediately and within the current patent system to address issues arising from the growth of patent applications. These strategies seek to improve the sharing of knowledge in relation to inventions and ultimately underpin a better patent system. While the patent system is based on the notion that a statutory right to control exploitation of the patented material is given in return for public disclosure as to how the invention works, access to and the ease with which one can understand such information is sub-optimal.

The first response is to build more accessible and meaningful patent databases including landscapes about patent portfolios in particular areas and what they cover. Being able to access and interpret worldwide patent information as easy as a Google search has the potential to vastly improve the innovation system. This can be greatly enhanced by the building of accessible patent landscapes that can provide researchers, governments and industry with vital information about what has been done and what can be done in the future. A leading initiative in this area is the "Patent Lens" (http://www.patentlens.net) a Web-based knowledge bank developed by Dr. Richard Jefferson and his institute CAMBIA.

The second response is to include peers in the patent application process. In much the same way as we use peers in academia to review quality a project has arisen in the United States called "Peer to Patent" (http://www. peertopatent.org) which embeds peers in the process of assessing patent applications. The Peer to Patent project was also trialed in Japan in 2008– 2009 and Australia in 2009–2010, and the concept is expected to spread to other patent offices throughout the world. In two years of operation in the United States, 187 applications have been subjected to the Peer to Patent process, which entails leaving the patent application open for public peer review for three months and then sending the top ten examples of prior art to the patent office for consideration in the examination process. Of the applications that have been subjected to peer review and then gone through to final determination approximately one in six have had "office actions" taken on the basis of the prior art revealed by the Peer to Patent process. The incentive for companies to participate (and big technology companies such as Microsoft, GE, IBM and Intel have participated) is that the US Patent Office (USPTO) will accelerate the assessment of the patent application by up to eighteen18 months if the applicant has participated in the Peer to Patent program.

There are four discernable benefits from such a project. Firstly, the applicant saves the time and expense of having a weak patent issued that will most likely be defeated at a later point in litigation or a patent issued that is robust which builds confidence in its use and exploitation. Secondly, industry competitors are incentivised at an early and relatively inexpensive stage to find prior art to allow a greater freedom to operate and lower the likelihood of conflict. Thirdly, there is a public benefit in the issuing of better-quality patents with a reduced likelihood of litigation. Fourthly, the USPTO benefits from being assisted by citizen experts (Obama and Biden 2008) and by not issuing non-meritorious patents that create inefficiencies within the innovation system. All of these benefits are produced by embedding the concept of open innovation into the administration process of one of the most critical aspects to an innovation system, namely, the examination of patent applications.

CONCLUSION

The premise of this chapter is that creativity and innovation will flourish in an environment where distributed and diverse ideas can be exchanged, experimented with and reused. Law has a critical role to play in this framework as its application and enforcement can have a major impact on the "flow of information", especially in the networked environment. As this chapter highlights, the choices we make in developing and interpreting the law are critical to how creative and innovative we can be as individuals, groups or firms and how productive—in a social, cultural and economic sense—Australia can be as a nation. Therefore, we must be mindful as to how the law operates and consider ways (such as those outlined here) in which it can be shaped to promote creativity and innovation.

NOTES

1. See generally: First Amendment to the US Constitution (United States); Art. 2 *Charter of Rights and Freedoms* (Canada); Art. 10 *The European Convention on Human Rights*.
2. "We—humankind—are a conversation. Because language is the medium in which reality is constituted, language is at once the house of being and home of human beings. Discourse is the literal translation of the Greek word *logos* which means *to make manifest or to let something be seen*" (Heidegger 1962, 55–56; emphasis added).
3. See *United States v. Microsoft Corp.*, 87 F. Supp. 2d 30; 97 F. Supp. 2d 59, 64–65 (D.D.C. 2000); 253 F.3d 34 (D.C. Cir. 2001).
4. Professor Tim Wu provides this definition: "Network neutrality is best defined as a network design principle. The idea is that a maximally useful public information network aspires to treat all content, sites, and platforms equally." See "Net Neutrality FAQ" http://timwu.org/network_neutrality.html.
5. "Net Neutrality Bill Introduced in Canadian Parliament" (2008).
6. All were pyrrhic victories in the sense that demand for file sharing has not abated and the companies that were pursued were inexperienced start-ups with limited cash reserves. This is not to say that these decisions did not slow

the widespread adoption of these new technologies or in a broader sense stop the fullest exploitation of these new technologies to our greatest benefit.

7. *Viacom International Inc. v. YouTube, Inc.*, 2007, S.D. N.Y., filed 13/3/2007. Viacom complaint available at http://graphics8.nytimes.com/packages/pdf/business/20100319-viacom-statement.pdf. YouTube and Google response available at http://news.com.com//pdf/ne/2007/070430_Google_Viacom.pdf.
8. See N. Weinstein's "Google Says Copyright Suit Threatens the Internet".
9. The Organisation for Economic Co-operation and Development (OECD) adopted the recommendations of the set of principles on *Open Access to Public Sector Information* at its Ministerial Conference in Korea in June 2008. See also OECD (2007); Fitzgerald, Coates and Lewis (2007); Bledsoe, Coates and Fitzgerald (2007); D. Bushell-Embling (2007); Queensland Spatial Information Council (2006); van Eechoud and B. van der Wal (2007).
10. See generally: *Campbell v. Acuff-Rose Music Inc.*, 510 US 569 (1994); Suzor (2006, 2008).
11. Green Day, "American Idiot", http://www.azlyrics.com/lyrics/greenday/americanidiot.html.
12. See *Gowers Review of Intellectual Property* (2006, 67–68); *Perfect 10 Inc. v. Amazon Com Inc.* 487 F.3d 701 (9th Cir, 2007).
13. Cf. BBC Creative Archive at http://www.bbc.co.uk/creativearchive.html.
14. See Queensland Spatial Information Council website: http://www.qsic.qld.gov.au/qsic/QSIC.nsf/CPByUNID/6C31063F945CD93B4A257096000CBA1A.

BIBLIOGRAPHY

Benkler, Y. 2006. *The wealth of networks: How social production transforms markets and freedom.* New Haven, CT: Yale University Press. http://www.benkler.org/Benkler_Wealth_Of_Networks.pdf.
Berman, P. S. 2000. Cyberspace and the state action debate: The cultural value of applying constitutional norms to "private" regulation. *University of Colorado Law Review* 71: 1263–1310.
Bledsoe, E., J. Coates and B. Fitzgerald. 2007. *Unlocking the potential through creative commons: An industry engagement and action agenda.* ARC Centre for Creative Industries and Innovation. http://creativecommons.org.au/unlockingthepotential.
Boyle, J. 1997. A politics of intellectual property: Environmentalism for the net? *Duke Law Journal* 47: 87–116. http://www.law.duke.edu/boylesite/Intprop.htm.
Bushell-Embling, D. 2007. Private eyes on public data. *Sydney Morning Herald*, 25 September. http://www.smh.com.au/news/technology/private-eyes-on-public-data/2007/09/24/1190486224755.html?page=fullpage.
Cerf, V. 2005. Vint Cerf speaks out on Net Neutrality. http://googleblog.blogspot.com/2005/11/vint-cerf-speaks-out-on-net-neutrality.html.
Chesborough, H. 2003. *Open innovation: The new imperative for creating and profiting from technology.* Boston: Harvard Business School Publishing.
Cutler, T. 2008a. "Innovation and open access to public sector information". In *Legal framework for e-research*, ed. B Fitzgerald, 25–39. Sydney: Sydney University Press.
———. 2008b. *Venturous Australia. Report of the Review of National Innovation System.* http://www.innovation.gov.au/innovationreview/Pages/home.aspx.
Dearne, K. 2008. Tanner eyes web 2.0 tools. *Australian IT*, 4 November. http://www.australianit.news.com.au/story/0,27574,24601440–15306,00.html.

Dicey, A. V. 1975. *An introduction to the study of the law of the constitution.* 10th ed. London: Macmillan.

Fisher, W. 2004. *Promises to keep: Technology, law and the future of entertainment.* Palo Alto, CA: Stanford University Press.

Fisher, W., M. Horwitz and T. Reed, eds. 1993. *American legal realism.* Oxford: Oxford University Press.

Fitzgerald, B. 1993. Proportionality and Australian constitutionalism. *University of Tasmania Law Review* 12: 263–322.

———. 1994. "Principles of Australian constitutionalism". Proceedings of the Australasian Law Teachers Conference, 19 September.

———. 2000. Software as discourse: The power of intellectual property in digital architecture. *Cardozo Journal of Arts and Entertainment Law Journal* 18: 337–86.

———, ed. 2007. *Cyberlaw Volume 1.* Aldershot: Ashgate.

———. 2008a. Copyright 2010: The future of copyright. *European Intellectual Property Review* 30: 43. http://eprints.qut.edu.au/archive/00013305.

———. 2008b. "Copyright 2010: The need for better negotiability/usability principles". In *Knowledge policy: Foresight for the twenty-first century*, ed. G. Hearn, D. Rooney and D. Wright. Cheltenham: Edward Elgar.

Fitzgerald, B., and B. Atkinson. 2008. "Third party copyright and public information infrastructure/registries: How much copyright tax must the public pay?" In *Knowledge policy for the twenty-first century*, ed. M. Perry and B. Fitzgerald. Toronto: Irwin Law.

Fitzgerald, B., J. Coates and S. Lewis, eds. 2007. *Open content licensing: Cultivating the Creative Commons.* Sydney: Sydney University Press.

Fitzgerald, B., and A. Fitzgerald. 2002. *Cyberlaw: Cases and materials on the Internet, digital intellectual property and e-commerce.* Sydney: LexisNexis.

Fitzgerald, B., A. Fitzgerald, G. Middleton, Y. F. Lim and T. Beale. 2007. *Internet and e-commerce law: Technology, law and policy.* Pyrmont: Lawbook Co., Australia.

Fitzgerald, B., A. Fitzgerald, M. Perry, S. Kiel-Chisholm, E. Driscoll, D. Thampapillai, and J. Coates. 2006. *Creating a legal framework for copyright management of open access within the Australian academic and research sectors. The OAK Law Project.* http://eprints.qut.edu.au/6099/.

Gowers Review of Intellectual Property. 2006. HMSO. http://www.hm-treasury.gov.uk/media/6/E/pbr06_gowers_report_755.pdf.

Graells, J., J. Soteras, and B. Verdejo. 2008. The use of Creative Commons licenses in the Ministry of Justice of the government of Catalonia. *Communia.* http://communia-project.eu/node/111.

Hartley, J. 2007. "The evolution of the creative industries—Creative clusters, creative citizens and social network markets". Keynote address to Creative Industries Conference, Asia-Pacific Forum, 19 September, Berlin.

Heidegger, M. 1962. *Being and time.* Trans. John Macquarrie and Edward Robinson. London: SCM Press.

Howkins, J. 2001. *The creative economy: How people make money from ideas.* London: Penguin.

Jefferson, R. 2006. Science as social enterprise: The CAMBIA BiOS Initiative. *Innovations* 1 (4): 13–44. http://www.bios.net/daisy/bios/3067/version/default/part/AttachmentData/data/INNOV0104_pp13-44_innovations-in-practice_jefferson.pdf.

———. 2007. *Freedom to innovate as a human right: The lost first pages.* http://blogs.cambia.org/raj/index.php/2007/04/05/freedom-to-innovate.

Jenkins, H. 2006. *Convergence culture—where old and new media collide.* New York: New York University Press.

Kelman, M. 1987. *A guide to critical legal studies*. Cambridge, MA: Harvard University Press.

Lauriault, T., and H. McGuire. 2008. *Data access in Canada: CivicAccess.ca*. http://www.osbr.ca/ojs/index.php/osbr/article/view/514.

Lemley, M., and L. Lessig. 2000. "The end of end-to-end: Preserving the architecture of the Internet in the broadband era". UC Berkeley Law & Econ. Research Paper No. 2000–19; Stanford Law & Economics Olin Working Paper No. 207; UC Berkeley Public Law Research Paper No. 37. http://papers.ssrn.com/sol3/papers.cfm?abstract_id=247737.

Lessig, L. 1999. *Code as law*. New York: Basic Books.

———. 2002. *The future of ideas*. New York: Random House. http://www.the-future-of-ideas.com.

Mayo, E., and T Steinberg. 2007. *The power of information (and UK government response)*. http://www.cabinetoffice.gov.uk/media/cabinetoffice/strategy/assets/power_information.pdf and http://www.cabinetoffice.gov.uk/media/cabinetoffice/strategy/assets/power_information_response.pdf.

Metcalfe, J. S. 2008. "The broken thread: Marshall, Schumpeter and Hayek on the evolution of capitalism". In *Marshall and Schumpeter on evolution, economic sociology of capitalist development*, ed. Yuichi Shinoya, 116–44. Northampton, MA: Edward Elgar.

"Net Neutrality Bill Introduced in Canadian Parliament". 2008. *Slashdot*. http://tech.slashdot.org/article.pl?sid=08/05/30/0133242.

Obama, B., and J Biden. 2008. *Science technology and innovation for a new generation*. http://www.barackobama.com/issues/technology/index_campaign.php.

O'Reilly, T. 2005. What is Web 2.0? *O'Reilly*. http://www.oreillynet.com/pub/a/oreilly/tim/news/2005/09/30/what-is-web-20.html.

Organisation for Economic Co-operation and Development. 2007. *OECD principles and guidelines for access to research data from public funding*. http://www.oecd.org/dataoecd/9/61/38500813.pdf.

Pearlman, J. 2007. Do-it-yourself charter to right future wrongs. *Sydney Morning Herald*, 1 December. http://www.smh.com.au/news/national/doityourself-charter-to-right-future-wrongs/2007/11/30/1196394622534.html.

Queensland Spatial Information Council. 2006. *Government information and open content licensing: An access and use strategy*. http://www.qsic.qld.gov.au/QSIC/QSIC.nsf/0/F82522D9F23F6F1C4A2572EA007D57A6/$FILE/Stage%202%20Final%20Report%20-%20PDF%20Format.pdf.

Schumpeter, J. 1943. *Capitalism, socialism and democracy*. London: Routledge.

Shapiro, C., and H. Varian. 1000. *Information rules: A strategic guide to the network economy*. Boston: Harvard Business School Press.

Suzor, N. 2006. "Transformative use of copyright material". LLM thesis, Queensland University of Technology.

———. 2008. Where the bloody hell does parody fit in Australian copyright law? *Media and Arts Law Review* 13: 218–48.

van Eechoud, M., and B. van der Wal. 2007. *Creative Commons licensing for public sector information: Opportunities and pitfalls*. Institute for Information Law, University of Amsterdam. http://www.ivir.nl.

Weinstein, N. 2008. Google says copyright suit threatens the Internet. http://www.zdnet.com.au/news/software/soa/Google-says-copyright-suit-threatens-the-Internet/0,130061733,339289364,00.htm.

Wunsch-Vincent, S., and G. Vickery. 2007. Participative Web and user-created content: Web 2.0, wikis and social networking. Organisation for Economic Co-operation and Development. http://www.oecd.org/dataoecd/57/14/38393115.pdf.

9   Towards a Sociology of Creativity

Janet Chan

INTRODUCTION

Creativity[1] has traditionally occupied a marginal position in sociology and, until the recent decades, few attempts have been made by sociologists to theorise creativity (Joas 1996; Domingues 2000). A review of the creativity literature readily demonstrates that the field has been dominated by philosophical and psychological, rather than sociological, analysis. This is not to say that the social aspects of creativity are ignored by researchers, but they often appear as vaguely defined variables such as "social environment" or "social influence" in a "shopping list" of individual, psychological and organisational factors associated with creativity. The construction of a *sociology* of creativity has not been given the attention it deserves.

This chapter is a first step in exploring the elements that can make up such a sociology. This project immediately raises a question: what is the value of a sociological framework for understanding creativity? To address this question, I will first examine ways of understanding creativity and then review the literature to identify various social dimensions that have emerged in previous research on creativity. It will be clear from this review that sociological approaches have a great deal to offer: the very meaning of creativity, the methods and processes of creative practice and the standards for judging creativity are all predominantly shaped by social structures and processes. Far from being peripheral to our understanding of creativity, sociology can help integrate the so-called "environmental" with the psychological factors to provide a more comprehensive and useful framework for studying creativity.

There is considerable challenge in formulating a sociology of creativity. Sociology is a contested and changing discipline (Giddens 1987); there is not a consensus on what sociology is and certainly there are many ways of doing sociology. It is therefore important to set out from the start that this chapter is about working towards a *particular* sociology of creativity; it is not an attempt to cover all available positions. The sociological model put forward will focus on creative action or practice, both at the individual and group levels, and consider the relationship between institutionalised structures and human agency. Such a model does not prejudge the meaning of creativity; rather, meanings are assumed to be fluid, emergent and socially constructed.

Fortunately, the project is already well under way. As Domingues points out, creativity has become a central theme for social theory since the turn of the century, although it has been "unfolding as a highly decentred theoretical movement, without common project or identity among its promoters" (2000, 468). The centrality of creativity as a theoretical concern is linked to a "new crisis of modernity":

> The heterogeneous character of contemporary social life, the rolling-back of the state and a greater reliance on society's own regulation, plus the crisis of the modernist imaginary, have led to a new situation in sociology wherein certainty has been replaced by uncertainty— and very often, we may add, by creativity—and contingency has to be recognised. (Domingues 2000, 479)

Although a number of theorists have touched on issues related to creativity (see Domingues 2000), two approaches—Joas's (1996) theory of creative action and Bourdieu's theory of practice—stand out as having laid the groundwork for creativity to be set in a sociological frame. Both theories will be examined, but Bourdieu's theory will emerge as providing a more useful framework for integrating the empirical findings of previous research and building an agenda for future research on creativity.

UNDERSTANDING CREATIVITY

What is creativity? In trying to define creativity, researchers have tended to make a distinction between "personal creativity" and "universal creativity", which separates less significant personal achievements from creativity that has a much greater historical significance and influence (Swede 1993; see also Boden 1994). Amabile specifies two "essential elements" in creativity: "A product or response will be judged as creative to the extent that (a) it is both a novel and appropriate, useful, correct or valuable response to the task at hand, and (b) the task is heuristic rather than algorithmic"[2] (1996, 35). Traditional definitions of creativity have typically focused on one or more of three aspects—the creative person, the creative process and the creative product[3] (Briskman 1981; Csikszentmihalyi 1996; Simonton 2003).

The Creative Person

Creativity is often connected with the "genius" who possesses imagination, understanding and reflective judgment. In his research based on nearly one hundred interviews with creative people in every field, Csikzentmihalyi concludes that creative people are different from non-creative people in their *complexity*: "they show tendencies of thought and action that in most people are segregated. They contain contradictory extremes—instead of being an 'individual,' each of them is a 'multitude'" (1996, 57). Amabile's (1996) review of the literature suggests that a limited set of personality traits has repeatedly been found to be associated with creative behaviour, including self-discipline, perseverance, independence, tolerance for ambiguity and willingness to take risks. Independence is one of the most consistent personality traits that have been associated with creativity. Amabile links this to intrinsic motivation, which has been found to be a "crucial determinant of creativity across different domains for different types of subject population" (1996, 115). Apart from possessing these personality traits, highly creative people are found to have certain cognitive abilities such as divergent thinking, including flexibility, originality, fluency and the ability to develop ideas (Watson 2007). They are also said to have "a flat hierarchy of associations" (as opposed to a steep hierarchy), which means that they are more able to make associations between disparate ideas (Simonton 2003).

The Creative Process

Creativity is often conceived as a process. Some have argued that there is a basic pattern in common to all creative processes (Koestler 1981). The creative process has been analysed in terms of stages, e.g. preparation, incubation, inspiration and elaboration, but such representations have been criticised as these stages are not necessarily linked to creativity (Glickman 1978). Authors have emphasised a certain amount of "controlled unpredictability" in the creative process. For example, artists exercise "critical judgment" to direct their creative activities, so that they are able to tell that "certain directions are not right", but the end result is not always predictable (Tomas 1979, 133). Thus, while there is "serendipity [in the creative act], it is not sheer accident" (Hausman 1981, 84–85). Koestler suggests that both scientific discovery and artistic inspiration involve a "temporary regression": scientists and artists are equally reliant on their "fallible intuitions" to take "a leap in the dark" (1981, 15). Briskman sums up the creative process as an iterative one of "critically interacting" with an established tradition and its problems, generating *"blindly, but not randomly"* a potential solution, then critically interacting with this initial product, and repeating this process until a product emerges that provides the sought-after solution or an improved solution (1981, 148).

Creative processes can vary between practitioners. Galenson's study of French and American painters identified two sets of practices that he labelled "seekers and finders": seekers' work is based on *perception*, they are engaged in an "extended process of searching for the elusive best

means" of presenting visual sensations; while finders' work is based on
conception, their goal being "to communicate their ideas or emotions" and
they engaged in presenting "a series of statements" (2001, 169). Drawing
on historical accounts of artists and their work practices, Galenson concludes that Cézanne was a typical seeker and Picasso a typical finder. The
two types of artists tend to work very differently: the seekers (the "experimental" artists) rarely make detailed plans and leave the most important
decisions to the working stage, whereas for the finders (the conceptual artists), detailed planning and preparatory studies are more important than
the execution of the plans. These differences in approach mean that experimental innovations often occur late in an artist's career, whereas conceptual innovations can occur at any age. In his more recent work, Galenson
(2006) extends this typology of seekers and finders to other creative artists including early painters, modern sculptors, poets, novelists and film
directors. He argues that his research sheds new light on the life cycle of
creative careers: "the pattern of creativity by age is not determined by the
activity, but by the approach of the creative individual" (2006, 176). More
important, his research challenges the assumptions of psychologists that
the "complexity of a field is exogenous to practitioners of that activity, and
that it is effectively immutable" (2006, 177). On the contrary, Galenson
finds that the complexity of a field can be radically changed by the practitioners: "innovators not only advance their disciplines, but they often
change them" (2006, 177).

The Creative Product

A number of authors have criticised the amount of attention that has been
paid to "the creative person" or "the creative process", when it is "the creativity of *the product* which has . . . logical priority" (Briskman 1981, 135).
Hemlin, Allwood and Martin also point out the obvious fact that "there is
no necessary connection between a creative production process and a creative product" (2004, 5). In art Tomas emphasises that the term "creative"
would not be applied to "any activity that does not result in a product having positive aesthetic or artistic value" (1979, 134). In the same vein, Best
argues, "A claim to be creative could not be justified by a mental experience
of a creative process, in the absence of a creative product" (1985, 82).

So what makes a product creative? Briskman lays down several criteria:
the product must be a "transcendent product" in that it is *novel*, it solves
a problem, does so in such a way as to "supplant or improve upon" parts
of the background of prior products; and it is "favourably evaluated" and
meets "certain exacting standards which are themselves part of the background it partially supplants" (1981, 144). Hemlin, Allwood and Martin
similarly define a creative product as "one that is not only novel and imaginative but also judged to be useful and of good quality" (2004, 4).

This definition of a creative product provides a link between creativity and innovation. Innovation can be defined as a useful product of creativity, be it a new procedure, design, system or object (Ramsey 2006; Rogers 2003). However, not all creative products are innovations. For example, a painting or sculpture that does not involve new methods or techniques of production may be a creative product but it would not be regarded as an innovation. Even "an idea for a new product or process" (an invention) is not an innovation; the latter requires the idea to be implemented in practice (Fagerberg 2005, 4).

Yet the judgment of a product as creative, as "new and valuable", as "original" and so on, often changes with time and depends on context. In a much quoted essay, Danto has argued that to "see something as art requires . . . an atmosphere of artistic theory, a knowledge of the history of art: an artworld" (1964, 580–81). This implies that artistic creativity is less about creating something new—mere novelty does not make art—than it is about creating something that is judged by the "aesthetic community" (artworld) to be original and valuable. Hemlin, Allwood and Martin similarly see the judgment of a product as "creative" as being "dependent on the prior knowledge and understanding brought into play" so that whether a product is considered creative or innovative may "vary between different contexts, and across different historical periods" (2004, 5). In art as well as in science, the novelty and value of a product has to be judged by the aesthetic or scientific community (Danto's [1964] "artworld" and Kuhn's [1962/1970] "paradigm"), although a consensus does not always exist within the community (Glickman 1978).

Deconstructing Creativity

If what is regarded as creative is context dependent, then the meaning of creativity may be contested, as it often has been in both artistic expressions and scientific discoveries. At a deeper level, our conception of creativity is based on a host of assumptions that are rarely examined. Rehn and De Cock have tried to unveil some of these assumptions by using Derrida's practice of deconstruction to "think creatively about creativity" (2009, 223). They suggest that deconstruction is one way of subjecting the concept of creativity to critique and questioning, in order "to empty the word of its dogmatic and ideological meanings" (2009, 229). Citing examples from art, design, technology and the underground economy of exchanges in the USSR, the authors argue that creativity need not always be defined by novelty or originality, nor seen as ethically neutral:

> By valorizing novelty over the pre-existing, one turns creativity into part of a modernist narrative of unending progress and the necessity of continuous capitalist development. By valorizing originality, one hides

away notions of production and work, not to mention history. By valorizing creativity as a neutral concept, one hides away the many assumptions about ethics and the nature of social life that form the possibility of normalizing concepts. We cannot fully escape the framework within which we think, nor the context from where we think, but we must work on our awareness about the foundations of our thinking. (Ibid.)

To formulate a useful sociology of creativity, it is important to be aware of the ideological or political assumptions underlying its definition, so that the concept can be fully captured and meaningfully explained.

SOCIAL DIMENSIONS OF CREATIVITY

It is possible to discern from the preceding discussion that while there may be psychological aspects of creativity relating to personality and motivation, social and cultural factors are in fact extremely important both for defining creativity and for understanding how it can be engendered. These factors are evident in the literature when discussing the social organisation of creativity and factors associated with the generation of creative outcomes, including innovation.

Social Organisation of Creativity

For a product to be regarded as creative, it has to be judged to be novel and valuable by a relevant community in a particular area of knowledge at a historical juncture. Csikzentmihalyi makes the distinction between a *domain*, which consists of shared cultural knowledge ("a set of symbolic rules and procedures"), and a *field*, which "includes all the individuals who act as gatekeepers to the domain" (1996, 27–28). It is the field that judges whether a new product is in fact valuable and novel in a significant way. But this judgment is not always based on some iron-clad, objective criteria that is time and culture invariant. Nor is there always consensus in the field regarding the criteria. As Hemlin, Allwood and Martin point out, this judgment is "usually a matter of social negotiation and thus at least partly influenced by rhetorical and other communication skills" (2004, 5). There are thus important social processes at work in the definition of creativity: the establishment of symbolic rules and procedures of a domain; the institutionalisation of judgment criteria used by the field; the negotiation of census regarding these criteria among members of the field (see Amabile 1996); the maintenance of stability of this judgment over time; the socialisation and education of creative practitioners to incorporate these rules, procedures and criteria into their practice; and the persuasion of consumers of the product to accept this judgment.

Socialisation into a Domain

In order to practise in a creative domain, a practitioner has to learn domain-relevant knowledge (Csikzentmihalyi 1996; Amabile 1996). For example, Best has argued that artistic creativity can be "educated" but only through grasping the criteria for assessing the creative product (1985, 80). The role of art education in developing creativity is thus grounded in teaching the technical, historical and theoretical knowledge required for an understanding of what counts as "good art". Similarly, Kuhn (1963, 343) argues that scientific "revolutions" are rare and "only investigations firmly rooted in the contemporary scientific tradition are likely to break that tradition and give rise to a new one". The teaching and learning of domain-relevant knowledge is one of the most important stages, if not *the* most important stage for the development of creativity.

Part of the learning also involves developing relevant skills and techniques for practising in a particular domain (Amabile 1996). Often these skills are not creative in themselves—they can be mundane and tedious, as in the practice of musical instruments or the mastery of technical skills in art making. As Kuhn (1962/1970) points out, at the level of everyday practice, "normal science" is not about innovation, but about puzzle-solving: it is through a deep engagement with the current tradition that scientists recognise an "anomaly" which requires further work that may eventually lead to a "paradigm shift". Although scientific revolutions are rare, discoveries or innovations of the "puzzle-solving" kind do occur regularly (see Knorr-Cetina 1981), as do artistic innovations of less dramatic nature.

Social Environment

The learning of domain-specific knowledge and skills and the practice of creativity do not happen in a social vacuum. Researchers such as Amabile (1996), Csikzentmihalyi (1996) and Hemlin, Allwood and Martin (2004) all acknowledge the importance of the social and work environment in shaping and developing creative practice. Being in the "right" environment is obviously important for accessing domain-relevant information, interactions and resources that stimulate creativity and facilitate the realisation of new ideas (Csikzentmihalyi 1996). More generally, "creative knowledge environments" (CKEs)—the physical, social or cognitive aspects of the environment—are important for engendering the production of new knowledge or innovations (Hemlin, Allwood and Martin 2004). CKEs can be as small as the environment of an individual worker or as large as the global environment of a multinational firm. Different types of CKEs are appropriate for different stages in the knowledge-production process: the problem-finding stage, the idea generation and elaboration stage, the evaluation stage and the "selling" stage (Hemlin, Allwood and Martin 2004).

Organisational Culture

Some researchers highlight the importance of organisational culture for engendering creativity (see also Mann, Chapter 10 this volume). Jarvie (1981), for example, suggests that two factors in the institutional arrangement—freedom to experiment and continuing encouragement—can make a difference to the development of creativity. Organisational theorists have suggested that culture is transmitted through a group-based learning process, either through positive reinforcement of successful solutions to problems ("problem-solving" learning) or through successful avoidance of painful situations ("anxiety-avoidance" learning) (Schein 1985). While problem-solving learning can engender innovations, anxiety-avoidance learning has the opposite effect. Thus, organisational leadership, incentive structures and management styles can enhance or dampen a culture of creativity. Benner and Tushman's (2002) research, based on a twenty-year longitudinal study of the paint and photography industries in the United States, suggests that process management activities widely adopted by organisations to increase efficiency may have "crowded out" technological exploration and innovation, making firms vulnerable at a time of rapid technological change.

Social Networks

While the dynamics of workers within teams or organisations have received a great deal of attention from researchers, a relatively neglected area is the role of social networks in facilitating creativity. Perry-Smith (2006) has argued that social interactions with other people can enhance a person's understanding of the domain as well as facilitate the learning of cognitive skills relevant for the generation of creative solutions. Her study of US researchers found that the number of *weak* social ties outside the organisation was positively associated with creativity but the number of strong ties was not. This was explained in terms of the heterogeneity of the background of the network, and hence the exposure to a variety of perspectives and ideas that may trigger a new approach. Contrary to expectations, the *centrality* of a person in a network had no direct impact on creativity. There appeared to be an *interaction* between centrality and number of outside ties: when a person had many outside ties, centrality made no difference to creativity, but the number of outside ties was positively associated with the creativity of a person who did not occupy a central position in a network.

In contrast with Perry-Smith's focus on network and creativity at the individual level, Ahuja's (2000) research examined network and innovation at the level of the firm. Using yearly patent counts of a firm as a measure of innovation, Ahuja analysed the number of direct and indirect ties as well as "structural holes" ("disconnections between a firm's partners") in ninety-seven chemical firms in several countries over a ten-year period. Controlling for a number of variables, such as R&D expenditure, firm size and so on, Ahuja found that the number of direct and indirect ties both had positive impact on innovation output, but the benefit from indirect ties decreased with the number of direct ties. The number of structural holes had a negative impact on innovation, even though it was hypothesised that the relationship could go in different directions depending on the context. Ahuja cautions against the building of large indirect networks to increase innovation even though they involve low maintenance costs, as the benefits of indirect networks may be relatively low and indirect ties may be competitors after the same information. In general, Ahuja concludes that the choice between dense or structural-hole-rich networks depends on the context as well as the type of benefits sought by the firm.

External Motivators

Csikzentmihalyi (1996) identifies reward as one of seven major social-environmental factors that contribute to creativity (the others are training, expectations, resources, recognition, hope and opportunity), although he only offers anecdotal evidence to support his thesis. The role of rewards or incentives has been the focus of Amabile's research for many years. In her updated book (Amabile 1996), the influence of external rewards on creativity is shown to be more complex than originally hypothesised. This led to a revision of the "Intrinsic Motivation Principle of Creativity" that she formulated in 1983. She defines as *intrinsic* "any motivation that arises from the individual's positive reaction to qualities of the task itself" and as *extrinsic* "any motivation that arises from sources outside of the task itself" (1996, 115). Her previous finding was that intrinsic motivation engenders creativity while extrinsic motivation is detrimental to it. Subsequent research has demonstrated that extrinsic motivation can also be conducive to creativity, but it has to be "in the service of intrinsics": "Rewards that confirm competence without connoting control, or rewards that enable the individual to do exciting work can serve as synergistic extrinsic motivators" (1996, 118). More generally, for Amabile, "social-environmental factors" can influence creativity positively or negatively depending whether they lead to *enabling* or *controlling* extrinsic motivations. Positive social environments are those that give practitioners autonomy, resources, encouragement and constructive feedback, while negative ones involve surveillance, control, win-lose competition among workers and critical evaluation that are threatening or humiliating.

Interdisciplinary Interactions

The building of interdisciplinary or multidisciplinary teams is another way of promoting creativity. Wilson (2002), for example, has argued that there is genuine opportunity for research breakthroughs when artists and scientists work together. This is because artists can bring valuable insights and perspectives to scientific research, given their "tradition of iconoclasm", their "valuing of creativity and innovation", as well as interest in social and communication aspects of science and technology (2002, 38). Contemporary examples of the collaboration between artists and scientists include various artist-in-residence programs in science and technology research centres (Wilson 2002; Harris 1999; Naimark 2003). For example, in the mid-1990s, the Palo Alto Research Center (PARC) of Xerox experimented with pairing new media artists with research scientists as a way of maintaining its innovative culture; the experiment led to a range of artist–scientist collaborations and some new ideas for scientific innovations (Harris 1999). More generally, a new domain of information technology and creative practices (ITCP) is already emerging (National Research Council 2003). Hollingsworth and Hollingsworth's research on 131 biomedical research organisations in the Untied States concludes that organisations that "have recurring major discoveries have tended to be those in which there is a high degree of interaction among scientists across diverse fields of science" (2000, 242). They explain the success of these organisations in the following way:

> The diversity of disciplines and depth of knowledge within a well-integrated research group or lab have the potential to change the way people view problems and to minimize their tendency to make mistakes and to work on trivial problems. . . . It is the diversity of disciplines and paradigms to which individuals are exposed with frequent and intense interaction that increases the tendency for creativity and for breakthroughs to occur. Working in an interdisciplinary environment without intense and frequent interaction among members of the work group does not lead to new ways of thinking (e.g., major discoveries). (2000, 243)

There is research evidence that new influences, such as new ideas, new contacts or new environments can be conducive to group creativity (Hemlin, Allwood and Martin 2004). Similarly, a change of field (domain of knowledge) can lead to creative advances through "combining different frames of reference or different preconceptions, or the transfer of a perspective, model or methodological approach from one discipline or subdiscipline into another" (Hemlin, Allwood and Martin 2004, 14–15; see also Dogan and Pahre 1990). Hemlin, Allwood and Martin suggest that researchers who switch fields are likely to be more creative because of their access to "unusual knowledge" which could "stimulate metaphorical thinking" (2004, 15).

SOCIOLOGICAL THEORIES OF CREATIVITY

The preceding review of the literature demonstrates that creativity is defined and recognised by relevant social groups, made possible through social learning and education and facilitated in particular by social environments and organisational settings. There is little doubt that a sociological approach to creativity is crucial for highlighting the significance of these social dimensions, but what would a useful sociology of creativity look like? Ideally, it should recognise that the definition of creativity is domain-specific and socially, culturally or politically constructed by the relevant community. A useful sociology of creativity should also account for variations in creativity in different domains, social groups, organisational units, geographical areas and historical epochs. It should provide conceptual tools for taking into account variations in skill, knowledge, motivation, access to social networks and resources and types of social environment. Finally, it should explain how creativity is possible by examining the relationship between social structure and human agency.

Amabile (1996) has argued that the lack of a general framework for a comprehensive psychology of creativity has led to a theoretical fragmentation within the field as well as a tendency for researchers to focus on narrow single-issue investigations. The same could be said about the sociology of creativity. As pointed out by Domingues, although creativity has "cropped up every now and then in sociological theory"—such as in Weber's notion of charismatic leadership, Durkheim's "collective effervescence", Marx's active and creative subject and so on—"the tendency to marginalise creativity was common to sociology in all the main national traditions" until recent years (Domingues 2000, 468). Among the social theorists who have approached issues related to creativity, two major theorists stand out as having made significant advances towards a useful sociology of creativity—the German sociologist Hans Joas and the French theorist Pierre Bourdieu. Their main theses will be examined briefly in the following. The utility of both approaches for understanding creativity has been examined in detail by Dalton (2004)—I will draw heavily on his essay in the following discussion.

Joas's Theory of Creative Action

Joas (1996) uses a sociological theory of action[4] to conceptualise creativity. He argues that in addition to the two predominant models of action—*rational* action and *normatively oriented* action—there should be a third, overarching model that "emphasizes the *creative* character of human action" (1996, 4). In fact, he asserts that "there is a creative dimension to all human action" (ibid.). Joas's theory of creative action takes issue with three tacit assumptions embedded in normative and rationalist models of action: first, that people are actually capable of taking "purposive action"; second, that they have control over their bodies; and, finally, that they are autonomous from other people and their environment. For Joas, normal action is habitual and "pre-reflexive": goals of action are usually not well defined, but actions take place within a social, historical and corporeal context through routinised habits. Creativity is what happens when these habits are "inter-rupted" and the actor succeeds in reconstructing the context through new ways of acting or thinking.

The flaw in Joas's framework, according to Dalton (2004), is that it sees creativity as a separate phase from habitual action. This implies that an action cannot be both creative and habitual, but such a framework would run into problems when explaining certain types of action. For example, when creativity is a routinised activity, such as in artistic practice, it is both a break with habitual action as well as a form of habitual action. For creative practice such as that among musicians, routinised action can be a "foundation for creative action" (2004, 609). In Dalton's view, Joas's framework "neglects the possibility that some action may contain creative and habitual elements simultaneously" (2004, 611). This brings us to a dis-cussion of an alternative, and much better-known theory developed by the French sociologist Pierre Bourdieu.

Bourdieu's Theory of Practice

Bourdieu's framework applies to social practice in general and is not explicitly about creativity, but he has used it to conceptualise the produc-tion of creative practices such as literary writing and visual art, as well as scientific research. The two key concepts *field*[5] and *habitus* are defined as follows:

> A field consists of a set of objective, historical relations between posi-tions anchored in certain forms of power (or capital), while habitus consists of a set of historical relations "deposited" within individual bodies in the form of mental and corporeal schemata of perception, appreciation and action. (Wacquant 1992, 16)

For Bourdieu, society is constituted by an ensemble of relatively autono-mous *fields*. A *field* is a social space of conflict and competition, where participants struggle to establish control over specific power and author-ity, and, in the course of the struggle, modify the structure of the *field* itself. Central to the concept of *field* is the notion of *capital*. There are vari-ous forms of capital which operate in different social *fields*. These include economic, cultural, social and symbolic capital, which is the form other types of capital take on when they are regarded as legitimate (Bourdieu 1987, 3–4). Bourdieu compares a *field* to a game where players possess tokens of different colours representing different types of capital.

In this framework, creative practice is the outcome of the interaction between a *field* of practice and the *habitus* appropriate to the creative *field*:

> the relationship between a creative artist and his work, and therefore his work itself is affected by the system of social relations within which

creation as an act of communication takes place, or to be more precise, by the position of the creative artist in the structure of the intellectual field (which is itself, in part at any rate, a function of his past work and the reception it has met). (Bourdieu 1969, 89)

Although the *field* generally shapes the *habitus*, the latter, to the extent that it is a "product of independent conditions", can have a relatively autonomous existence and in turn help shape the *field* (Bourdieu 1993, 61).

When conceptualising creativity, then, it is more appropriate to think in terms of a specific *field* of creative practice, such as fine arts, literature, design, science or technology. Within each field, there is a configuration of *capital*, such as prestige, status, wealth, power or knowledge, that provides a structural dimension that recognises the resources as well as constraints of practising in that field. Each practitioner in the field occupies a particular position, depending on his or her accumulated capital. The scientific field, for example, is structured by various types of interests, such as the acquisition of "scientific authority"—a form of capital which combines both intellectual (or technical) and political capital (Bourdieu 1975). Inevitably, every "scientific choice"—the selection of problem, methods, publication venues, etc.—is geared towards the "maximisation of strictly scientific profit, *i.e.* of potential recognition by the agent's competitor-peers" (Bourdieu 1975, 22–23; Stephan 1996). Because of the race to establish priority of discovery, scientists compete fiercely in "winner-takes-all contests" to be the first to publish (Stephan 1996, 1202).

The artistic field is a site of constant struggle between two ordering principles: the *heteronomous principle* of popular art, which values the "bottom line" or economic capital, and the *autonomous principle* of "high" art, which values artistic independence, art for art's sake and avant-garde practice (Bourdieu 1993, 40). There is also the struggle between "positions", for example, between "the orthodoxy of established traditions and the heretical challenge of new modes of cultural practice, manifested as . . . position-takings" (Johnson 1993, 16–17). Galenson provides a vivid account of what he calls "intergenerational conflict", which "has been a distinct feature of modern art" (2001, 155). Within the artistic *field*, what counts as valuable or legitimate ("art" as opposed to "non-art") is the product of such struggles (Bourdieu 1993, 261).

In Bourdieu's theory, *habitus* is a system of "dispositions" which integrate past experience and enable individuals to cope with a diversity of unforeseen situations (Wacquant 1992)—dispositions which agents acquire either individually, through family and the education system or as a group, through organisational socialisation. *Habitus* generates strategies which are coherent and systematic, but it also allows for creation and innovation within the field of creative practice. It is a "feel for the game"; it enables an infinite number of "moves" to be made in an infinite number of situations.

The field also produces a belief or investment in the game—what Bourdieu calls *illusio*, which accepts that the game is worth playing or worth being taken seriously. In fact, the existence of the game is both the cause and the effect of the "permanent production and reproduction of the *illusio*, the collective adherence to the game" (Bourdieu 1996, 167).

Scientists' *habitus*, for example, is developed through university education and workplace socialisation. The position of individual scientists in the scientific *field* and in their career trajectory may affect whether they adopt a "succession strategy" which does not challenge accepted orthodoxies or a "subversion strategy" which does (Bourdieu 1975, 30). Ironically, it is those rich in scientific capital who are more likely to have the ability to bring about a "scientific revolution" (1975, 33).

For an artist, the habitus would consist of assumptions about what art is about, what is considered good art and ways of "doing art" that have proved to be useful. For a creative artist, the "game" is to produce works of art that are judged to be good art by the artistic community, and so there is usually not a conscious effort to produce *creative work* as such. Artists typically develop their artistic habitus initially in art schools, where they acquire knowledge of artistic techniques, art history and art theories. Apart from gaining technical competence in their area of specialisation, art students also learn ways of seeing, thinking and working, as well as attitudes and sensibilities that constitute the artistic habitus. In other words, they learn the aesthetic judgment that can help "direct" the process of their own art practice (Tomas 1979). Fine art students are more likely to adopt the values of authenticity and independence (autonomous principle) than those of populism and commercialism (heteronomous principle). Galenson's research suggests that external conditions (the *field*) can "foster shifts in the prevailing conception of art" (2001, 165).

Explaining Creativity

Although Bourdieu never explicitly set out to provide a theory of creativity, Dalton suggests that the definition of *habitus* provides for the possibility of creative agency. This happens at two levels. The first source of creativity flows from the "inevitably imprecise fit between general dispositions and concrete situations"—the *habitus* is "necessarily plastic, permitting innovation in the carrying forth of specific actions" (Dalton 2004, 613, 612). The second source of creativity lies in the strategic actions taken by individuals within a *field* that is oriented towards creativity: practitioners are motivated by the status or recognition they could achieve by working according to the dispositions which are shaped by the *field*. Dalton is concerned, however, that this offers only a "restricted realm of creativity" and therefore does not provide for a "robust conception of creative agency":

Bourdieu does not seem to allow here for the type of thoroughgoing reconstruction of principles, goals, or methods that a fully rational or creative actor hypothetically could produce. For Bourdieu, creativity is always a restricted set of strategies embedded in the bodily hexis and the logic of a particular social milieu . . . Habitus is flexible and open-ended but continues to place significant bounds on the "horizons of possibilities" that Joas describes. (2004, 613)

Dalton's suggested solution is to "sever the connection emphasized in Bourdieu between habit and received cultural and social patterns" and construct a new model of creativity that combines the best of Joas's and Bourdieu's models by reaffirming Joas's idea that "creativity is an inherent feature of action that exists within both highly routinized activities and within more self-evidently creative conduct" (2004, 615). In other words, the necessity to innovate is part of an overarching *habitus* that people share, regardless of their *field* of practice. When faced with obstacles and difficulties, people may follow the habituated way of dealing with problems allowed in their *field* or they may question or challenge these accepted ways and act in a creative way that transcends or transform their *field*. This is consistent with Kuhn's (1962/1970) account of the "scientific revolution" that requires a "paradigm shift".

One advantage of this reformulation of Bourdieu's framework is that it makes it possible to distinguish routinised activities in a creative field from innovative activities that have both the potential to redefine the field and still be recognised as creative within the "creative community".

CONCLUSION

A review of the literature has highlighted the importance of the social dimensions of creativity. In this chapter, these social dimensions are examined not simply as a bundle of isolated variables to be manipulated, but considered in a framework that goes some way towards a coherent and useful sociology of creativity. Bourdieu's theory of practice—with some minor modifications suggested by Dalton (2004)—provides the ingredients as well as the architecture for such a framework. The strengths of this framework can be summarised as follows.

First, consistent with the general creativity literature, the framework explicitly defines creativity in terms of specific *fields* (or Csikzentmihalyi's domains), each of which involves field-specific *habitus* (knowledge, skills or a "feel of the game"). Within each *field*, standards or criteria for judging creativity are determined by the community of practitioners, experts, consumers and/or markets. Organisations within a *field* can be regarded as *subfields* with their own sets of power relations and "games".

Second, consistent with research findings that access to knowledge, support, resources and communities that engender creativity are unevenly distributed, the notion of *capital* in Bourdieu's framework provides a useful way of conceptualising this inequality. Creative practitioners occupy certain social positions in the *field* according to the types and amounts of capital they have accumulated. Each *field* has its own definition of what types of capital are valued and what weights are attached to them. For example, autonomy as a symbolic capital is particularly salient among non-commercial artists and scientists, while experience and skills are valued as cultural capital among designers and engineers.

Third, in line with findings of research on organisational or work group culture that encouragement, freedom and a climate of trust are conducive to creativity, Bourdieu's theory postulates that practice—in this case creative practice—is the outcome of the interaction between the *subfield* and the *habitus*. A skilled worker may be intrinsically motivated to innovate, but the organisational structure and culture may be detrimental to creativity.

Finally, the framework, especially as modified by Dalton (2004), allows for creative breakthroughs when practitioners overcome difficulties resulting from a change of environment or domain, working in a multidisciplinary team, or occupying a position of marginality. When a worker, who has been socialised in a particular *field* or *subfield* and as a result acquired a particular *habitus*, changes to a new *field*, he or she would inevitably experience discomfort and unease. Bourdieu's theory suggests that the worker's *habitus* will adjust incrementally to fit with the new field (Bourdieu 2000). Such an adjustment may produce creativity as a matter of course through cumulative incremental work, or as a paradigm shift under certain conditions. Similarly, multidisciplinary teams involve people with different *habituses* working together, thus increasing the opportunity for the combination of different frames of reference or transfer of disciplinary perspectives or methodologies that may lead to creative outcomes.

Note that although Bourdieu's framework is primarily sociological, the notion of *habitus* is in fact socio-psychological-biological, as it consists of assumptions, dispositions and modes of practice embodied in an individual through socialisation into a *field*.

Although I have justified the framework with reference to the existing research literature, the utility of a sociology of creativity is not simply to summarise what we already know, but to provide a structure for conceptualising future research and further refinement of the theory. Empirical research on creative practitioners and their *field* of practice can shed light on a variety of research questions: e.g. How is creativity understood and practised in different *fields*? How do variations in resources and constraints (dimensions of capital) in the *field* or *subfield* affect the practice of creativity? What impact do changes in the *field* have on the *habitus* of creative practitioners and the judgment criteria used by creative communities?

What happens when creative practitioners from two or more *fields* are put to work in a combined setting? These are all important questions that can enrich the field of creativity research and open it up to broader conceptual questions that are currently not tackled by creativity research studies.

ACKNOWLEDGMENT

I would like to thank Noreen Metcalfe for her valuable assistance with the collation and annotation of the research literature. Comments from Leon Mann, Mark Dodgson and Kerry Thomas on an earlier version of this chapter are much appreciated.

NOTES

1. This chapter focuses on creativity and only briefly on innovation. Innovation will be discussed as a particular type of creative *product*, one that has been successfully implemented or adopted in practice.
2. Heuristic tasks differ from algorithmic ones in that the former do not have a "clear and readily identifiable path to solution" whereas the latter do (Amabile 1996, 35).
3. Watson (2007) also listed "press, place and persuasion" as additional perspectives on creativity.
4. Joas discusses Talcott Parsons's theory of action at great length but he does not see Parsons's normativist conception of action as the best way to go beyond the rational model.
5. Bourdieu's concept of *field* will be italicised to avoid confusion with Csikzentmihalyi's field, which represents gate-keepers of the domain.

BIBLIOGRAPHY

Ahuja, G. 2000. Collaboration networks, structural holes, and innovation: A longitudinal study. *Administrative Science Quarterly* 45 (3): 425–557.

Amabile, T. M. 1996. *Creativity in context: Update to the social psychology of creativity.* Boulder, CO: Westview Press.

Benner, M. J., and M. Tushman. 2002. Process management and technological innovation: A longitudinal study of the photography and paint industries. *Administrative Science Quarterly* 47: 676–706.

Best, D. 1985. *Feeling and reason in the arts.* London: Allen and Unwin.

Boden, M. A. 1994. *Dimensions of creativity.* Cambridge, MA: MIT Press.

Bourdieu, P. 1969. Intellectual field and creative project. *Social Science Information* 8(2): 89–119.

———. 1975. The specificity of the scientific field and the social conditions of the progress of reason. *Social Science Information* 14 (6): 19–47.

———. 1987. What makes a social class? On the theoretical and practical existence of groups. *Berkeley Journal of Sociology* 32: 1–18.

———. 1993. *The field of cultural production.* Cambridge: Polity Press.

———. 1996. *The rules of art.* Cambridge: Polity Press.

————. 2000. *Pascalian meditation*. Cambridge: Polity Press.

Bourdieu, P., and L. J. D. Wacquant. 1992. *An invitation to reflexive sociology*. Cambridge: Polity Press.

Briskman, L. 1981. "Creative product and creative process in science and art". In *The concept of creativity in science and art*, ed. D. Dutton and M. Krausz, 129–56. The Hague: Martinus Nijhoff Publishers.

Csikszentmihalyi, M. 1996. *Creativity: Flow and the psychology of discovery and invention*. New York: Harper Collins.

Dalton, B. 2004. Creativity, habit, and the social products of creative action: Revising Joas, incorporating Bourdieu. *Sociological Theory* 22 (4): 603–22.

Danto, A. 1964. The artworld. *Journal of Philosophy* LXI: 571–84.

Dogan, M., and R. Pahre. 1990. *Creative marginality: Innovation at the intersections of social sciences*. Boulder, CO: Westview Press.

Domingues, J. M. 2000. Creativity and master trends in contemporary sociological theory. *European Journal of Social Theory* 3 (4): 467–84.

Fagerberg, J. 2005. "Innovation: A guide to the literature". In *The Oxford handbook of innovation*, ed. J. Fagerberg, D. C. Mowery and R. R. Nelson, 1–26. Oxford: Oxford University Press.

Galenson, D. W. 2001. *Painting outside the lines*. Cambridge, MA: Harvard University Press.

————. 2006. *Old masters and young geniuses: The two life cycles of artistic creativity*. Princeton, NJ: Princeton University Press.

Giddens, A. 1987. *Social theory and modern sociology*. Stanford, CA: Stanford University Press.

Glickman, J. 1978. "Creativity in the arts". In *Philosophy looks at the arts: Contemporary readings in aesthetics*, rev. ed., ed. J Margolis, 143–61. Philadelphia: Temple University Press.

Harris, C., ed. 1999. *In search of innovation*. Cambridge, MA: MIT Press.

Hausman, C. R. 1981. "Criteria of creativity". In *The concept of creativity in science and art*, ed. D. Dutton and M. Krausz, 75–90. The Hague: Martinus Nijhoff.

Hemlin, S., C. M. Allwood and B. R. Martin, eds. 2004. *Creative knowledge environments: The influences on creativity in research and innovation*. Cheltenham: Edward Elgar.

Hollingsworth, R., and E. J. Hollingsworth. 2000. "Major discoveries and biomedical research organizations: Perspectives on interdisciplinarity, nurturing leadership, and integrated structure and cultures". In *Practising interdisciplinarity*, ed. P. Weingart and N. Stehr, 215–44. Toronto: University of Toronto Press.

Jarvie, I. C. 1981. "The rationality of creativity". In *The concept of creativity in science and art*, ed. D. Dutton and M. Krausz, 109–28. The Hague: Martinus Nijhoff.

Joas, H. J. 1996. *The creativity of action*. Trans. J. Gaines and P. Keast. Chicago: University of Chicago Press.

Johnson, R. 1993. Editor's introduction: Pierre Bourdieu on art, literature and culture. In *The Field of Cultural Production*, by P. Bourdieu, 1–25. Cambridge: Polity Press.

Knorr-Cetina, K. D. 1981. *The manufacture of knowledge: An essay on the constructivist and contextual nature of science*. Oxford: Pergamon Press.

Koestler, A. 1981. "The three domains of creativity". In *The concept of creativity in science and art*, ed. D. Dutton and M. Krausz, 1–17. The Hague: Martinus Nijhoff.

Kuhn, T. S. 1962/1970. *The structure of scientific revolutions*. Chicago: University of Chicago Press.

———. 1963. "The essential tension: Tradition and innovation in scientific research". In *Scientific creativity: Its recognition and development*, ed. C. W. Taylor and F. Barron, 341–354. New York: Wiley.

Naimark, M. 2003. *Truth, beauty, freedom, and money: Technology-based art and the dynamics of sustainability: A report for Leonardo Journal.* www.art-slab.net.

National Research Council. 2003. *Beyond productivity: Information technology, innovation, and creativity.* Washington, DC: The National Academies Press.

Perry-Smith, J. E. 2006. Social yet creative: The role of social relationships in facilitating individual creativity. *Academy of Management Journal* 49 (1): 85–101.

Ramsey, R. D. 2006. The supervision of innovation. *Supervision* 67 (7): 3–5.

Rehn, A., and C. De Cock. 2009. "Deconstructing creativity". In *The Routledge companion to creativity*, ed. T. Richards, M. A. Runco and S. Moger, 222–31. New York: Routledge.

Rogers, E. M. 2003. *Diffusion of innovations.* New York: Free Press.

Schein, E. 1985. *Organizational culture and leadership.* San Francisco: Jossey-Bass.

Simonton, D. K. 2003. Scientific creativity as constrained stochastic behavior: The integration of product, person, and process perspectives. *Psychological Bulletin* 129 (4): 475–94.

Stephan, P. E. 1996. The economics of science. *Journal of Economic Literature* 34(3): 1199–1235.

Swede, G. 1993. *Creativity: A new psychology.* Toronto: Wall and Emerson.

Tomas, V. 1979. "Creativity in art". In *Art and philosophy: Readings in aesthetics*, 2nd ed., ed. W. Kennick, 131–42. New York: St. Martin's Press.

Wacquant, L. J. D. 1992. "Toward a social praxeology: The structure and logic of Bourdieu's sociology". In *An invitation to reflexive sociology*, ed. P. Bourdieu and L. J. D. Wacquant, 1–59. Cambridge: Polity Press.

Watson, E. 2007. Who or what creates? A conceptual framework for social creativity. *Human Resource Development Review* 6 (4): 419–41.

Wilson, S. 2002. *Information arts: Intersections of art, science, and technology.* Cambridge, MA: MIT Press.

10 Social Psychology of Creativity and Innovation

Leon Mann

INTRODUCTION

The activities of creativity and innovation occur in a social context. In some instances creative people, ostensibly working alone, benefit from sponsorship, social support and recognition. In other cases, creativity stems from people working in pairs, teams, partnerships and networks. Innovation, the realisation and utilisation of new ideas and concepts in practice, is also by nature a social phenomenon, as it depends on users, backers and investors to support and adopt the new idea. For creativity to be realised as innovation, there must be implementation and usage by others who see the benefit, value or intrinsic merit of the new idea, concept or product. Thus both creativity and innovation are social phenomena.

Teresa Amabile (1996), a social psychologist, regards creativity as a necessary (but not sufficient) condition for innovation. "Creativity by individuals and teams is a starting point for innovation" (Amabile et al. 1996, 1155). Amabile brings creativity and innovation together in the context of organisations. She somewhat narrowly defines innovation as the successful implementation of creative ideas within an organisation. But it is an entirely valid point that organisations such as research institutes, universities, company R&D laboratories and design centres are the settings in which a great deal of creativity is fostered and innovation planned. In many respects creativity is a larger and more challenging activity than innovation. Creativity encompasses the activity of generating novel ideas and thinking irrespective of purpose and setting and goes beyond individuals and teams working in organisations. In many modern economies, creativity, the essential prerequisite for innovation, tends to be assumed or neglected while the catchphrase "innovation" dominates public policy, business thinking and education and training as the key to national performance and competitiveness.

The main messages conveyed in this chapter are: first, the significance of the research team as the basic organisational unit for driving creativity and innovation in advanced industrial societies; second, the striking importance of the so-called "soft" factors of leadership, trust, positive team dynamics, organisational support and encouragement, as keys to the performance of all teams in the quest for creativity and innovation; and third, the importance of attention to the design and composition of research teams and to the training of early career researchers in leadership and collaboration skills to increase the potential for creativity and innovation.

SOCIAL PSYCHOLOGY AND CREATIVITY/INNOVATION

The field of social psychology is concerned with social influence, social motivation, interpersonal relationships, communication, leadership and group behaviour, group processes (such as conformity, compliance and conflict resolution) and organisational dynamics and performance. The social psychology of creativity deals with such topics as intrinsic motivation and social rewards as incentives to creativity (Amabile 1996), social support that assists creativity (consider the Medici's patronage of Michelangelo, Raphael and Donatello), the relationship between creative duos (consider Rogers and Hammerstein in music, Watson and Crick in physiology, Pierre and Marie Curie in radiology), the dynamics of creative teams (for example, Lockheed's "Skunk Works" in the 1990s) and the climate of creative organisations (consider IDEO, IBM, Xerox PARC and Optiscan).

Numerous case studies of highly creative individuals, duos, groups and organisations provide evidence of the importance of social psychological factors in creative endeavours and innovation. The focus shifts from the study of creativity towards the study of innovation as the level of analysis moves from the individual and pair to the team, large group and the organisation.

Howard Gardner, in *Creating Minds* (1993), comments on the significance of affective and cognitive support from significant "others" (in this case, close friends, associates or family members) as a factor that helps creative individuals to make major breakthroughs. He illustrates the importance of social support in his study of the lives of seven highly creative people: Freud, Einstein, Picasso, Stravinsky, T. S. Eliot, Martha Graham and Gandhi. In the case of creative pairs or duos, for example, Pierre and Marie Curie, the role of intimacy, inspiration, complementary talents, sharing ideas and candid feedback is recognised. Chadwick and Courtivron's (1993) *Significant Others* examines creativity from the perspective of unions between creative partners in artistic and literary fields, for example, Diego Rivera and Frida Kahlo, and August Rodin and Camille Claudel. At the level of the large group, Bennis and Biederman (1997), in *Organising Genius: The Secrets of Creative Collaboration*, analyse social factors in their examination of seven creative groups (some more creative than others), for example, the Walt Disney Studio (animated feature films), Xerox PARC (personal computers), Lockheed's Skunk Works (radically new planes) and the Manhattan Project (the atomic bomb). The authors point to strong leadership, highly talented people with considerable autonomy, a keen sense of mission and self-perception as underdogs against a common "enemy" (i.e. competitors and rivals) among the factors associated with highly creative teams.

Peters and Austin (1985), in *A Passion for Excellence*, were among the first to identify social factors in innovative organisations such as Hewlett-Packard and 3M (for a discussion of innovative companies, see also Dodgson, this volume). Based on their observations, Peters and Austin formulated a model for successful innovation: an organisational climate that nurtures and recognises champions in all phases of innovation—from discovery, through development, to marketing; a champion (skunk) who takes up an idea and pushes it to fruition; "skunk works"—a small team of creative mavericks who persist because they have ownership and feel committed to the idea; and a belief in constant experimentation and testing. Note that this analysis of innovation in organisations points to three connected levels: individuals (champions/skunks), teams (skunk works) and organisations (a climate conducive to innovation and also recognition by management).

A FOCUS ON THE TEAM

Teams are incubators of creativity and innovation (Lipman-Blumen and Leavitt 1999). Increasingly the team is seen as the key to creative work in the world of science and technology and in the world of cinema, design, music and the arts (installations, collaborations). One study in the comic book industry found evidence that simply working in a team can, under the right circumstances, produce more creative results than working individually (Taylor and Greve 2006). On average, single creators had lower performance than did teams, and the team experience of working together increased performance.

Teams are important for scientific creativity and technological innovation because the very nature and complexity of the projects and problems tackled by today's researchers necessitate drawing on the expertise of diverse professionals from many disciplines to find the best solution or pooling the talents and time of many people to achieve a rapid solution. In an era of the knowledge society and the knowledge organisation there is growing interest in the characteristics of highly creative teams (Lipman-Blumen and Leavitt 1999; Hackman 2002; Mann 2005). This interest is reflected in the attention given by major research organisations to team design and training. An example is Australia's largest research organisation, CSIRO (Mann and Marshall 2007). The interest is also reflected in the establishment of research partnerships and alliances and networked organisations to provide the critical mass and expertise to tackle significant problems and undertake major projects (for example, mapping the human genome).

While some creative "teams" exist outside the framework of an organisation or institution, e.g. a group of neighbours who make a home-made rocket in their garage, this phenomenon is rare. Most creative teams exist within the framework of an organisation and are established and resourced by organisations, funding agencies and sometimes venture capital.

Research teams have as their *raison d'être* the discovery and application of knowledge. Examples of research teams include a team searching for a solution to why fish are dying in a river system, a team searching for a cure for Alzheimer's disease, a team working on a new way of measuring clinical depression and a team developing an inexpensive battery for electric automobiles. Our focus is on the team as the key source of creative and innovative endeavour in modern organisations, including university and company laboratories, media and design companies, architecture and engineering firms, think-tanks and public policy analysis institutes. This is not to underestimate the significance of the truly creative individual, the "bright spark" whose original creative insights drive the team and organisation—consider Walt Disney and Bill Gates as examples. But much—probably most—creativity is due to the combined but often unrecognised efforts of many individuals who contribute their ideas and opinions.

The research team is the basic unit of modern research organisations. Research teams are established because the questions tackled in research organisations and the products developed in company R&D laboratories require the knowledge and contribution of specialists and experts, sometimes from a range of disciplines (e.g. scientists, engineers, geographers, statisticians) and from a range of functions (e.g. product development, operations, sales, marketing). Mann and Marshall (2007) observe that another reason for research teams is the important role of *tacit* knowledge as a trigger to creativity and innovation. Tacit knowledge is the insight and understanding gained by team members from their immediate experience, hunches and intuition, trial and error, tweaking old procedures and talking informally with colleagues and clients. Tacit knowledge, unlike *explicit* knowledge, is rarely documented. Indeed, people are often unaware of their own tacit knowledge until they are questioned or challenged by others. Mann and Marshall conclude, "Teams are by far the most effective way of uncovering and sharing tacit knowledge, largely because such knowledge often emerges informally in face-to-face interactions, casual chats between colleagues, and in team meetings" (2007, 139).

Research on Research Teams: The ARC "Success Factors in R&D" Project

I will draw upon work by myself and colleagues at the University of Melbourne to discuss a social psychological approach to the study of creativity and innovation in research teams. From 1996 to 2000 I was involved in an Australian Research Council (ARC) longitudinal study of fifty-eight research and development (R&D) teams in four large Australian research organisations (Mann 2005). The organisations were CSIRO and DSTO (two public research organisations), BHP (a mining company) and Orica (a chemicals and paints company). Each team was studied for over twelve months during which time they were surveyed regularly on their team climate, communication patterns, evaluation of their leader, etc., and measured on performance (e.g. customer evaluations, patent applications, technical reports, etc.). The teams were followed up six months later to check for additional indicators of innovation, such as new products. The aim of the longitudinal study was to examine the impacts of leader, team and organisational variables on team innovation and performance over time.

While all fifty-eight teams in the ARC study were R&D teams, they differed in purpose. Some were pure research teams conducting fundamental research to create new knowledge; others were product development teams, extending and applying existing knowledge to improve existing products and technical processes (cf. Leifer and Triscari 1987). The distinction between research teams and development teams is important for understanding team creativity and innovation, because research teams usually have greater scope and expectation to perform highly creative work that *might* (or might not) lead to a major discovery or innovation, while development teams aim for an incremental improvement or refinement to an existing product or process.

Project M at BHP is an example of a highly creative research team in the ARC study. Project M produced a radical innovation—a revolutionary new technology for casting super-thin strips from molten steel—and it produced seven patents and three patents expected, new and improved processes and products and a host of technical and refereed publications. Orica's Genuine Enamel Mimic (GEM) project is an example of a creative product development team in the ARC study. The GEM team produced a new range of highly successful washable paints (Paints 101) for the Australian market and had one patent and six technical reports. Creativity is essential for both research teams and product development teams. Lisa Madigan, team leader of the GEM project observed:

> In the sort of work we were doing, innovation came into . . . the problem-solving because it was a results-driven project with goals to achieve. We hit pretty heavy obstacles along the way that might not have been solved using conventional problem-solving approaches. So we had to really think outside the box and get to our solutions really quickly or else the project time-lines would have been shot.

Team Climate for Innovation

Good teams are made up of team members who have different views, ideas and areas of expertise for generating new approaches and knowledge (West,

Borrill and Unsworth 1998). Creative teams have an ethos that places the highest value on solving task-related and technical problems and persisting until the objective is achieved. This ethos is especially important in successful research teams. But bringing together a group of diverse people and asking them to produce a brilliant new concept or solve a difficult problem does not lead to innovation if they cannot work well together, or if the project goals are vague or devalued. A team environment that supports innovation, allows creative ideas to be openly communicated, fairly evaluated and properly implemented is required (Amabile and Gryskiewicz 1987).

Team climate refers to the atmosphere and culture of the team, the norms and standards of behaviour, the mood of goodwill or hostility that pervades team meetings and how members work together. Michael West (1990) postulated four conditions that characterise a climate of innovation in teams: *Participative Safety*, an open atmosphere of discussion and participation, so that all ideas can be openly presented and discussed; *Support for Innovation*, team members help each other to pursue innovative goals, such as giving practical advice and sharing resources; *Objectives* or *Vision*, teams have clear and important goals that all team members find acceptable and motivating; and *Task Orientation*, team members have a common concern to meet team objectives at the highest level. The four factors are the basis of the Team Climate Inventory (TCI) (Anderson and West 1998).

Team climate for innovation has been studied in a wide range of settings, including work teams (Agrell and Gustafson 1994), business management teams (Burningham and West 1995), senior management teams in hospitals (West and Anderson 1996) and research teams (Bain, Mann and Pirola-Merlo 2001). Teams scoring well on the TCI have greater self-reported and actual innovation. Lisa Madigan, leader of the GEM team, commented about team climate:

> In my team we have regular fairly free-speaking group meetings . . . People update each other on where they are going and what their actions are and if they have any specific problems. That's a forum for us to discuss [those problems]. I might chair, but basically it's an open discussion and everyone is encouraged to offer ideas to others because they are working in similar veins.

What creates a positive team climate for innovation? Much of it comes from the goodwill, trust and talent of the team members themselves, but it is also due to the guidance of a capable team leader, especially in the areas of knowledge building and sharing (see the following). Strong organisational encouragement for innovation, in the form of senior management support for innovation and open discussion of ideas, is also important.

The team's climate for innovation is important in R&D teams, but we predicted that team climate would be more important for the performance of research teams than development teams. Our hypothesis was tested using data from the fifty-eight teams in the ARC study who were administered the TCI and measured on six indicators of innovation, including new products and processes. In accordance with our assumption, team climate was more strongly and consistently related to innovativeness in research teams than in development teams. We found that the TCI factor "support for innovation" (team members help each other to achieve innovative goals by giving practical advice and sharing resources) was the strongest correlate of team innovativeness and innovations in research teams, e.g. number of patents ($r = .37$). The correlations were weaker in development teams. The ARC study provides evidence of a connection between team climate and the reported and actual innovativeness of R&D teams.

Creative Individuals and Innovative Teams

The question of whether a team is only as creative as its most creative member (or "brightest spark") is of interest as it is relevant to matters of team selection and design and also the question of what makes some teams more creative than others. Is it a case of one or two highly gifted team members doing all or most of the creative work or is it a case that everyone's contribution counts? There are without doubt numerous examples to support both viewpoints. My own view is that everyone's contribution counts. Pirola-Merlo and Mann (2004) examined the relationship between individual innovativeness and team innovation using data from the ARC study. For nine months team members rated how innovative (i.e. creative) they had been in the previous month from 0 ("not at all innovative") to 10 ("highly innovative"). Then nine months later the team leader reported the number of new products, processes, patents and patent applications resulting from the project. To be sure, some of these new products and processes were well underway in the period when the scientists were rating their own innovativeness. But in many cases, the innovative products and outcomes had yet to see the light of day. As expected, individual innovativeness ratings fluctuated across time in keeping with the phase of project work and the obstacles and setbacks experienced. On average, team members rated their innovativeness at 5 out of 10 with a range of 2 to 7 (Pirola-Merlo 2000). It is striking that the monthly aggregate of individual innovativeness ratings turned out to be a reasonably good indicator of actual innovations recorded for the team many months later. The average correlation between monthly individual innovativeness ratings (aggregated to the team level) and number of innovations later documented for the team was $r = .23$ for new products/processes and $r = .38$ for new patents and patent applications. Two messages can be taken from these findings: first, that the innovativeness (creativity) of many individuals combines to produce the team's innovation in products and outcomes; second, that every team member's contribution counts—it all adds up.

RESEARCH ORGANISATIONS AND
ORGANISATIONAL SUPPORTS FOR INNOVATION

As noted earlier, creative research in science and technology is conducted not by lone individuals but by teams in research organisations, whether universities, research institutes, company labs or government agencies. Thomas Edison considered his best invention to be his "invention factory" (his laboratory at Menlo Park, New Jersey, established 1876) because it enabled him to create all his other inventions (Mattimore 1993). In his laboratories Edison employed teams of workers to systematically investigate a given subject. Edison boasted he could turn out a minor invention every ten days and "a big thing every six months or so" (Buderi 2000). Indeed Edison's "labs" gave birth to more than thirteen hundred US and foreign patents. Edison's laboratories were the forerunner of the modern industrial research laboratories established in the twentieth century by major companies such as Bell Labs, General Electric, DuPont, Westinghouse, Procter and Gamble and Xerox to create new product ideas and concepts in chemicals, pharmaceuticals, household products, electrical goods, telecommunications and so on.

The great industrial R&D companies turned to social and organisational psychology for ideas about the best structure and design for their R&D function. One persistent question was whether to establish a single central corporate research laboratory or separate research division or whether to have a cluster of small, specialised research groups attached to separate business units (see also West, Chapter 12, this volume). The benefit of the central corporate lab is pooled resources, availability of major facilities, broad expertise and opportunities for knowledge sharing between different units for strategic, long-term research and innovation (O'Connor and DeMartino 2006). The benefit of separate labs in each business unit is close contact between researchers, managers and customers to anticipate and respond to specific and local needs in product development. At the time of the ARC study (1996–2000), BHP had a central corporate lab, Orica had three specialist labs and CSIRO and DSTO had separate labs for each of their many divisions. The question of organisational structure and design for the R&D function is bound up with the issue of the company's mission and whether its goal is radical innovation, incremental innovation or perhaps a mixture of the two (Christensen 1997).

Organisational Support for Innovation: From
Team Climate to Organisational Climate

Beyond the issue of organisational structure and design, there is the matter of sustaining an organisational culture conducive to creative teams and innovation (see Peters and Austin 1985). Amabile's (1997) componential model of creativity specifies three components of individual and small team creativity: expertise, creativity and intrinsic task motivation. Amabile extended the model to encompass the social environment in organisations and how it might impact on individual and team creativity. Together with colleagues she developed an instrument for assessing the climate for creativity in organisations known as KEYS, and began studying project teams in a company dubbed "High Tech Electronics International". Experts in the company rated the company's research and development projects on creativity ("the production of novel and useful ideas by individuals of teams"). She found that compared with the low-creativity projects, the high-creativity projects enjoyed much greater environmental stimulants to creativity—positive challenge, freedom, supervisory encouragement, work group supports, organisational encouragement and sufficient resources.

Pirola-Merlo and Mann (2005) built on Amabile's work in their study of how organisational support for innovation influences the amount of innovation produced by the fifty-eight teams in the ARC study. The sum of patents/patent applications, publications and new and improved products was used as an index of team innovation. An instrument known as Organisational Supports for Innovation Questionnaire (OSIQ; Pirola-Merlo 2000) was used to measure three aspects of organisational climate as reported by team members: *encouragement* of innovation—does the organisation seem to value and encourage innovation?; *resources*—is there adequate provision of facilities, materials, information, time and access to experts?; and *empowerment*—do team members feel encouraged and have the autonomy to develop new ideas? The three organisational climate factors correlated significantly with measures of team innovation. The strongest factor was organisational encouragement—the perception that the organisation expects and supports innovation. Pirola-Merlo and Mann (2005) report that team climate for innovation and organisational encouragement of innovation each make a unique contribution to team innovation and together account for nearly 50 per cent of the variance in explaining team innovation. In sum, team and organisational climate are significant predictors of team innovation. The message is that the so-called soft factors of team climate and organisational encouragement make a difference and are crucial for producing hard innovation outcomes.

The Leadership Factor

It has been said that behind every creative team is a competent team leader. This section describes some findings from the ARC study about the significance of team leadership style, role performance and trustworthiness for team climate and creative performance.

Transformational leaders inspire a sense of mission about the importance of the team's work and stimulate new ways of thinking and problem-solving. Howell and Avolio (1993) describe how transformational leaders in the business units of a financial institution created a team climate conducive to creative thinking and novel solutions. Gillespie and Mann

(2005) measured transformational leadership in the sample of fifty-eight team leaders in the ARC study and examined whether it was related to team climate. Indeed, transformational leadership correlated highly with innovative team climate in teams led by new leaders ($r = .77$) and those led by experienced leaders ($r = .67$). Thus leaders highly regarded for challenging and inspiring their team make a significant difference to the climate for innovation in their teams.

Trust in the leader, which relates to their competence, reliability and willingness to share information, is another factor responsible for a positive team climate for innovation. Gillespie and Mann (2005) analysed data from the ARC study and found that team members' belief in the trustworthiness of their leader was significantly correlated with all four team climate factors, especially participative safety and task orientation. Gillespie and Mann also found that the leader's trustworthiness has a dynamic effect on team climate. Increases in leader trustworthiness over time were followed months later by improved team climate and performance.

Bain and Mann (1997) identified four leadership roles that are important in research teams. They are knowledge builder (provides scientific and technical knowledge to the team), stakeholder liaison (coordinates the team's task with key stakeholders and customers), standards upholder (maintains proper procedures and priorities) and team builder (builds a positive team climate and sound relationships between team members). Bain et al. (2005) measured the performance of the fifty-eight team leaders in the ARC study on the four leadership roles. They found that team leaders at the helm of teams most highly evaluated by project sponsors and customers were those held in highest regard for their capabilities as knowledge builders, stakeholder liaisons and team builders by their colleagues. In a follow-up study of engineering project teams in a large automotive company, Lee et al. (2010) examined the dynamics of the leader's role as knowledge builder. Team leaders who were highly regarded as knowledge builders had a positive effect on trust and sharing of task-related knowledge between team members, and this in turn had a significant effect on team performance in project work. The correlations between the leader's reputation as a knowledge builder and the team's knowledge sharing and innovation performance were significant. To conclude, the characteristics and quality of the team leader make a difference to the team climate for innovation, knowledge sharing and team performance.

The Creativity–Innovation Pipeline

There is a path from creativity to innovation. The flow of ideas within and into the team provides a pool from which the most novel and useful ideas are selected by the sponsoring organisation for further review and possible support and adoption. Companies now invest considerable resources in "ideas pipelines", casting their net widely even across regions and industries for novel and interesting ideas to turn into product and process innovations. Research partnerships, alliances, collaborations, university research centres and so on are all part of the ideas pipeline for finding novel ideas and concepts for new product development (cf. Procter and Gamble, Rio Tinto and other companies). Ideas are at the heart of team creativity and accordingly innovative organisations. What innovative organisations attempt to achieve are: first, conditions favourable to the generation of numerous ideas by teams (a kind of collective "brainstorming"); second, conditions favourable to generating highly unusual, sometimes implausible, but just possibly brilliant ideas that open up new ways of looking at things; and third, conditions favourable to the refinement and selection of a subset of ideas to be taken further for concept testing and development *en route* to possible innovation.

We have discussed team climate—especially psychological safety—as conducive to creative thinking and innovative outcomes. Other social psychological processes relevant to the functioning of creative research teams include freedom from conformity pressures; openness to the possibility of minority viewpoints and influence; avoidance of what Irving Janis (1972) identified as "groupthink" tendencies (illusions of unanimity and invulnerability, self-censorship and "mind-guards"); awareness of the tendencies toward polarisation or extreme group opinions and positions; attention to "runaway" norms, i.e. tendencies to try to outdo others in expressing group values—especially in a risky direction, in-group insularity and the "not invented here" (NIH) syndrome; the Abilene Paradox ("no one really liked the idea but everyone went along for the ride"); and commitment traps ("we've invested so much time and resources already, we can't quit now"). This writer has observed several research teams and knows of many others where dysfunctional group processes derailed the work of the team and sapped its creativity.

Also relevant to the discussion of dysfunctional group processes are factors in the organisation that interfere with the objective selection of a subset of creative ideas to take forward for possible implementation or commercialisation. While idea selection within the organisation is supposed to be a systematic, logical process, the selection of project ideas and concepts to take forward to possible commercialisation or implementation is often in practice a messy business plagued by a myriad of social and "political" pressures, including perseverance by the most vocal advocates, power and status plays and coalition formation to advance special parochial interests.

Team Structure and Composition

In focusing on teams—and in particular research teams—I have touched on several team process variables relevant to creativity and innovation, such as team climate, knowledge building and sharing, trust and team leadership. There are also a number of team structure and composition variables relevant to team innovation; they include team size, member diversity, team duration, team stage in its life cycle and virtual teams.

Team size. What is the ideal size of a research team to foster creativity? Of course, it is simpler to establish a strong team climate for innovation, knowledge sharing, trust, etc., in small than in large teams. But Stewart (2006) points out that large teams have an advantage, as presumably they have sufficient numbers to provide a wide array of knowledge, skills and expertise to find creative solutions to difficult, complex projects. Some organisations design their project teams to capture the autonomy and creativity of small teams within the structure of a larger team. For example, Pfizer Inc. organises its projects into "families" of five to seven scientists and then gathers about ten "families" into a "tribe" of seventy people (Wilmert 2001).

Team member diversity. Real teams (as distinct from groups) are those whose members are committed to a common purpose, set of common goals and working approach (Katzenbach and Smith 1993). The more diverse the team membership, the greater the likelihood of divergent approaches and viewpoints, which is presumably beneficial for creativity. The question arises: how much diversity and of what kind? A review of the literature by Mannix and Neale (2005) suggests that team diversity can just as easily lead to negative as to positive outcomes. They reviewed fifty years of research on teams and concluded that the preponderance of evidence yields a pessimistic view: group diversity creates social divisions, with negative performance consequences. The authors suggest that positive effects, such as creativity, can arise from differences in functional background, education or personality—but only when the group process is managed carefully. In another review of the literature, Hülsheger, Anderson and Salgado (2009) made an important distinction between *job-relevant diversity*, i.e. the range of attributes, such as profession, function, skills and expertise, and *background diversity*, i.e. the range of differences in age, gender and ethnicity. In a meta-analysis of studies of team innovation they found that job-relevant diversity is positively related to team innovation, while background diversity is negatively related. The overall message is that diversity can be good for creativity but it depends on the kind of diversity and in all cases must be managed well.

Team duration. Some research teams run out of creative steam when the same members have worked together for a long time and are starved of new ideas and fresh perspectives. There is evidence that with increasing tenure R&D project teams display lower levels of communication with experts outside their own project group or organisation (Katz 1982). The question arises of how best to introduce new ideas and ways of doing things in veteran teams that are becoming stale and complacent? This problem did not occur in the research teams in the ARC study, probably because they had at least some membership turnover during their project life cycle. Adding and changing team members and bringing in new members "on secondment" are ways of maintaining fresh thinking in teams.

Team stage. All research teams go through a "life cycle" from start-up to finish. There are more or less intensive bursts of creativity at different phases. Pirola-Merlo (2000) observed from his analysis of research teams in the ARC study that the planning stages of a project require more creative, divergent thinking, while the routine action phases require more procedural convergent thinking. It is important for managers to understand the stage of the creativity–innovation cycle when they impose pressures and demands on a team. In some circumstances, a sense of urgency and time pressure can be beneficial for creativity (Hennessey and Amabile 2010). But putting a team under time pressure to quickly settle on an approach to tackle a complex problem can be detrimental to creativity in the early stages of a project when interest should be to get all ideas on the table and discuss them thoroughly before "plunging in".

Virtual teams. In teams where members meet regularly there is considerable informal sharing of tacit knowledge which is highly beneficial for creative thinking and problem-solving. This feature is missing in most virtual teams. The question is how to share tacit knowledge between team members working in virtual teams in different locations. One answer is to arrange for virtual teams to meet face-to-face occasionally with time provided for the informal exchange of ideas. Another is for the team leader to be a regular "boundary rider" to each location and engage each team member in informal discussions.

CONCLUSION AND SOME RECOMMENDATIONS

In this chapter we have taken a social psychology perspective to examine creativity and innovation, focusing on creative teams within innovative research organisations. We have focused on teams and their leaders, but also on the role of organisations where the teams do their work. Creativity and innovation go hand in hand in modern research organisations. Supposedly, the connection between creativity and innovation is linear (a creative useful idea is implemented and adopted as a new innovative product or technology). However, this is an idealised picture of the relationship between creativity and innovation (See Mann, Chapter 15, this volume, for a discussion of multiple pathways linking creativity and innovation). Many highly creative and useful ideas are not implemented as innovations or wait for a more favourable environment for their adoption and implementation.

The research team is one link in the chain between creativity and innovation. The links include, at one end, the talented individuals who are selected to join the team and, at the other end, the organisation itself (including leaders and managers), which provides the encouragement, resources and autonomy for teams to put forward new ideas and solutions to consider for innovation. A key to examining the connection between creativity and innovation in organisations is to use multilevel analysis to examine individual and team factors (creativity focus) and organisation factors (innovation focus) and how they interact and fit together. Organisations which place a high value on radical innovation are likely to emphasise careful recruitment, selection, training and leadership of project teams to stimulate creativity. Senior management in those organisations provide their teams with considerable autonomy and incentives to experiment and think outside the box. Conversely, the most creative teams tend to use their reputation and influence within the organisation to garner resources and also management support to promote their ideas and move them toward adoption and implementation (see especially the team leader's role in stakeholder liaison). Sometimes tension arises between the administrative constraints and procedural requirements of the organisation and the freedom and spontaneity valued by highly creative teams (cf. Katzenbach and Smith 1993). But the most creative teams are found in the most innovation-minded organisations—and the relationship flows from organisation to team and back again.

In conclusion, here are some recommendations for government, industry and public research organisations (including universities) based on the analysis in this chapter. First, government must provide adequate support and incentives for companies and public-sector organisations to engage in fundamental (high-risk, creative) research with high potential for major innovation. Government should reduce administrative and bureaucratic constraints that create disincentives to expenditure on industrial R&D and to high-risk, high-value fundamental research in public research organisations. Second, all organisations—but especially research organisations—should monitor whether they are doing enough to encourage and support team creativity and innovation A good starting point is for organisations to audit the three dimensions identified earlier as organisational supports for innovation: encouragement of innovation, resources and empowerment. Finally, universities and public research organisations should train and develop all postgraduate research students and early career researchers in the skills of team leadership and how to work collaboratively in teams. As we have seen, so-called "people skills" of transformational leadership, trustworthiness, communication, knowledge sharing, etc., make a significant difference to team creativity and innovative performance. Learning and development of these important skills should not be left to chance or to the "school of hard knocks". They should be a core part of the education and professional socialisation of every researcher.

BIBLIOGRAPHY

Agrell, A., and R. Gustafson. 1994. The team climate inventory (TCI) and group innovation: A psychometric test on a Swedish sample of work groups. *Journal of Occupational and Organizational Psychology* 67: 143–51.

Amabile, T. 1996. *Creativity in context*. Boulder, CO: Westview Press.

———. 1997. Motivating creativity in organisations. *California Management Review* 40 (1) 39–58.

Amabile, T. M., R. Conti, H. Coon, J. Lazenby and M. Herron. 1996. Assessing the work environment for creativity. *Academy of Management Review* 39 (5): 1154–84.

Amabile, T., and S. S. Gryskiewicz. 1987. *Creativity in the R&D laboratory. Technical Report Number 30*. Greensboro, NC: Center for Creative Leadership.

Anderson, N. R., and M. A. West. 1994. *The team climate inventory*. Windsor: ASE/NFER-Nelson Press.

———. 1998. Measuring climate for work group innovation: Development and validation of the team climate inventory. *Journal of Organizational Behavior* 19: 235–58.

Bain, P. G., and L. Mann. 1997. *The project leadership questionnaire: Validation and preliminary norms. Melbourne Business School working paper series*. Melbourne: University of Melbourne.

Bain, P. G., L. Mann, L. Atkins and J. Dunning. 2005. "R&D project leaders: Roles and responsibilities". In *Leadership, management, and innovation in R&D project teams*, ed. L. Mann, 49–70. Westport, CT: Praeger.

Bain, P. G, L. Mann and A. Pirola-Merlo. 2001. The innovation imperative: The relationship between team climate, innovation and performance in research and development teams. *Small Group Research* 32 (1): 55–73.

Bennis, W., and P. W. Biederman. 1997. *Organising genius: The secrets of creative collaboration*. Reading, MA: Addison-Wesley.

Buderi, R. 2000. *Engines of tomorrow: How the world's best companies are using their research labs to win the future*. New York: Simon and Schuster.

Burningham, C., and M. A. West. 1995. Individual, climate, and group interaction processes as predictors of work team innovation. *Small Group Research* 26 (1): 106–17.

Chadwick, W., and I. de Courtivron, eds. 1993. *Significant others: Creativity and intimate partnership*. London: Thames and Hudson.

Christensen, C. M. 1997. *The innovator's dilemma: When new technologies cause great firms to fail*. Boston: Harvard Business School Press.

Gardner, H. 1993. *Creating minds*. New York: Basic Books.

Gillespie, N., and L. Mann. 2005. "How trustworthy is your leader? Implications for leadership, team climate, and outcomes in R&D teams". In *Leadership, management, and innovation in R&D project teams*, ed. L. Mann, 93–122. Westport, CT: Praeger.

Hackman, J. R. 2002. *Leading teams: Setting the stage for great performance*. Boston: Harvard Business School Press.

Hennessey, B., and T. Amabile. 2010. Creativity. *Annual Review of Psychology* 61: 569–98.

Howell, J., and B. Avolio. 1993. Transformational leadership, transactional leadership, locus of control and support for innovation. *Journal of Applied Psychology* 78 (6): 891–902.

Hülsheger, U. R., N. Anderson and J. F. Salgado. 2009. Team-level predictors of innovation at work: A comprehensive meta-analysis spanning three decades of research. *Journal of Applied Psychology* 94 (5):1128–45.

Janis, I. 1972. *Victims of groupthink: Psychological studies of policy decisions and fiascos.* Boston: Houghton Mifflin.

Katz, R. 1982. The effects of group longevity on project communication and performance. *Administrative Science Quarterly* 27: 81–104.

Katzenbach, J. R., and D. K. Smith. 1993. *The wisdom of teams.* New York: Harper Collins.

Lee, P., N. Gillespie, L. Mann and A. J. Wearing. 2010. Leadership and trust: Their effect on knowledge sharing and team performance. *Management Learning* 41(4): 473–491.

Leifer, R., and T. Triscari Jr. 1987. Research versus development: Differences and similarities. *IEEE Transactions on Engineering Management* 34: 71–78.

Lipman-Blumen, J., and H. J. Leavitt. 1999. *Hot groups: Seeding them, feeding them, and using them to ignite your organization.* New York: Oxford University Press.

Mann, L., ed. 2005. *Leadership, management, and innovation in R&D project teams.* Westport, CT: Praeger.

Mann, L., and R. J. Marshall. 2007. Teams in CSIRO: Reorganising for national research imperatives. *Innovation: Management, Policy & Practice* 9 (2): 136–45.

Mannix, E., and M. Neale. 2005. What differences make a difference? The promise and reality of diverse teams in organizations. *Psychology in the Public Interest* 6: 31–55.

Mattimore, B. W. 1993. *99% Inspiration.* New York: AMACOM.

O'Connor, G., and R. DeMartino. 2006. Organising for radical innovation. *Journal of Product and Innovation Management* (November): 475–97.

Peters, T., and N. Austin. 1985. *A passion for excellence.* New York: Random House.

Pirola-Merlo, A. 2000. "Innovation in R&D project teams: Modelling the effects of individual, team and organisational factors". PhD thesis, University of Melbourne.

Pirola-Merlo, A., and L. Mann. 2004. The relationship between individual creativity and team creativity: aggregating across people and time. *Journal of Organisational Behaviour* 25 (2): 235–57.

———. 2005. "Organisational supports for innovative R&D". In *Leadership, management, and innovation in R&D project teams*, ed. L. Mann, 211–30. Westport, CT: Praeger.

Stewart, G. L. 2006. A meta-analytic review of relationships between team design features and team performance. *Journal of Management* 32: 29–55.

Taylor A., and H. R. Greve. 2006. Superman or the fantastic four? Knowledge combination and experience in innovative teams. *Academy of Management Journal* 49: 723–40.

West, M. A. 1990. "The social psychology of innovation in groups". In *Innovation and creativity at work: Psychological and organizational strategies*, ed. M. A. West and J. L. Farr, 309–33. Chichester: John Wiley and Sons.

West, M. A., and N. R. Anderson. 1996. Innovation in top management teams. *Journal of Applied Psychology* 81 (6): 680–93.

West, M. A., C. S. Borrill and K. L. Unsworth. 1998. "Team effectiveness in organisations". In *International review of industrial organisational psychology*, ed. C. I. Cooper and I. T. Robertson, 1–48. Chichester: John Wiley.

Wilmert, T. 2001. Smart workplace design should bring people and their ideas together. *Minneapolis Star Tribune*, 28 January.

11 Creativity and Innovation Management
Play's the Thing

Mark Dodgson

Creativity and innovation are both issues of considerable interest to the management field as they are the primary means by which organisations survive and thrive in changing and uncertain circumstances. Study of the management of creativity and innovation, however, has evolved as two separate fields: scholars consider creativity or innovation and rarely the relationships between them. As creativity is commonly seen as the origination of ideas and insights and innovation as their application to commercial advantage in new products, services and business models, then the disjuncture from a management perspective is somewhat surprising. It would appear sensible for the study of management to examine any practices or behaviours that assist the connections between new ideas and their application. This chapter examines some issues for management theory and practice surrounding the nexus between creativity and innovation, provides data revealing how these links are not being examined in the academic literature and suggests that the concept of play around the use of new design and virtual reality technologies may help build the association between these two distinct strands of understanding. A number of short case studies are provided to illustrate the significance of playing with technology at work.

AT FIRST SIGHT: SOME ASPECTS OF THE RELATIONSHIP BETWEEN CREATIVITY AND INNOVATION MANAGEMENT

To begin with the most obvious questions: why and where is the relationship between creativity and innovation management important and how is it manifested? For private-sector managers, the reason for the importance of the relationship is simple: it delivers economic value and competitive advantage, their raison d'être. The "why" question has to be posed in the context of an economy that values knowledge as a source of intellectual property and operational expertise and flexibility in fast-evolving, uncertain and complex business environments. In a Schumpeterian or evolutionary economics framework, creativity is a source of entrepreneurial variety from which selection and propagation mechanisms establish innovations and produce the "creative destruction" characteristic of economic development. For Schumpeter (1942/1975), innovation is defined by the new combinations of the economic system's supplies of productive means and the Schumpeterian entrepreneurs who stimulate the recombination of resources undertake an essentially creative act.

The management of the relationship between creativity and innovation significantly affects the nature of work. Marx wrote about how the attractiveness of work results from the free play of physical and mental powers (Marx 1976, 284). His analysis of the capitalist labour process, of course, concluded that creative acts become separate from their implementation at work, and Marxist analyses contend managerial strategies accentuate this separation (Braverman 1974). Yet many contemporary businesses see the encouragement of creativity within an organisation as a means of making work more attractive, improving the engagement and commitment of existing staff and a winning strategy in the "war for talent" amongst highly skilled and mobile employees (Florida 2002). Creativity is argued to be core to the strategic development of the firm (Amabile 1996, 1998; Bilton 2007).

Margaret Boden tells us that "[c]reativity is a feature of Everyman" and is "based in ordinary abilities that we all share, and in practiced expertise to which we can all aspire" (1990, 245–46). Much of the management literature on creativity has tended to focus on individuals or the role of team members (Belbin 1993; McFadzean 1998; Puccio 1999), addressing techniques to extract the best performance from individuals, such as creativity training (Snow and Couger 1996), conflict resolution (Thomas 1983) and monetary rewards (Chee and Phuong 1996).

Indeed, as might be expected in the management field, with its common association with consulting opportunities, there is no shortage of tools and techniques on offer designed to increase the creative performance of organisational members. These range from Edward De Bono's role-playing "6 coloured hats", to notions of "mind maps" and more formally structured approaches such as "Ideation" and "TRIZ". Few, if any, of these tools and techniques are troubled in their promotion by a complete lack of any empirical evidence of their efficacy. Nonetheless, the eagerness with which they are received and disturbingly high levels of profit flowing to their originators show the continuing enthusiasm of organisations to promote the creativity of Everyman.

Essentially, the literature on the management of creativity has tended to address issues related to the creativity of individual and teams. It is generally highly dependent upon theories from within psychology and organisational behaviour.

The management of innovation, by contrast, generally has as its focus the organisation and the broad range of contextual factors that contribute to its development. Since the early research of Burns and Stalker (1961) and Woodward (1965), the primary concern has been with understanding the relationships between market and technological contexts, organisational structures and innovation. This continues with contemporary interests in the nature and objectives of innovation strategies, the management of research and development, how market and customer information is elicited and used, how partnerships with external parties are constructed and the internal organisational structures and practices that facilitate effective knowledge transfer and use (Dodgson, Gann and Salter 2008). Just as there is no management theory of creativity, there is no theory of the management of innovation, although there are some useful insights provided by the strategic management literature in fields such as transaction costs economics, the resource-based view and dynamic capabilities theory (see Helfat et al. 2007), which are themselves often reliant on theories from economics and industrial organisation.

Bilton (2007) decries the creativity literature's past lack of attention to context, and persuasively argues creativity needs to be considered within the systems and strategies that shape its manifestation. Indeed, his definition of creativity as something that is new, valuable and useful is almost identical to a common definition of innovation as the "the successful application of new ideas". However, as he acknowledges, his approach engages much more with the creativity rather than the innovation literature, and there remain many opportunities for improved understanding by building closer links between the two.

The innovation literature distinguishes between various *levels* and *types* of innovation and the relationship between creativity and innovation depends upon which is addressed and sought. Greater creativity is obviously associated with more radical innovations, although there is always a requirement for new ideas and insights in the search for continuous, incremental innovations. Innovations in products, services and processes are more common than in business models (i.e. strategies for delivering value). The relationship between innovation and creativity varies with *sources* of activity. When innovation is sought from customers and suppliers, the level of creativity present is usually significantly less than if innovation is sought from university research. The nexus between creativity and innovation can be found in: *individuals, teams, organisational divisions* and *firms* (with continuing questions about the comparative advantages of large firms with substantial resources versus the flexibility and responsiveness of small firms, and how these may be combined). It can be found in *social groups*, with the notion of the "creative class" (Florida 2002); *alliances*, with consideration of issues such as "open innovation" (Chesbrough 2003); *networks, clusters, sectors* and *regions*, especially allied to issues around creative or cultural industry and notions of the "innovative milieu" (Cooke and Morgan 2000; Caves 2000; Breschi and Malerba 2005; Potts 2009); and *nations*, with continuing debate, for example, about the imitative rather than creative industrial performance of many Asian nations (Kim 1997; Sigurdson 2005).

The advantages of combining insights from the innovation and creativity literatures can be seen in a wide range of managerial concerns. The perennial question for organisation and innovation and creativity, for example, is balancing the tension identified by Jim March (1991) of the need to simultaneously "exploit" and "explore". The management literature proposes the idea of the "ambidextrous" organisation, with parts focusing on operational efficiency in established areas of business, and others seeking new business and technological opportunities. The separation of two distinct groups with different structures, incentives and cultures within the same organisation is fraught with difficulties, and many companies have experimented with isolated teams (skunk works), semi-sanctioned periods of individual experimentation (bootlegging) and fully sanctioned routines allowing "free" time for the pursuit of innovative projects (Dodgson, Gann and Salter 2008).

Challenges abound in managing the balance of innovative teams: the need for established working practices and procedures, organisational learning and trust, versus the need for team refreshment and dissonant voices. Robert Sutton (2006) castigates organisations that ignore his "no asshole" rule, but admits that difficult colleagues and nasty managers can occasionally be organisationally stimulating. Similarly, within networks, to overcome the problems of institutional isomorphism—the more organisations work together, the more they come to resemble one another (DiMaggio and Powell 1983)—there is a need to work with dissimilar organisations. Hence the "collaboration paradox": the need to partner with organisations that are the most difficult to collaborate with (Dodgson 1993). Firms that work too closely with their customers and suppliers have been shown to become blind to the dangers and opportunities of disruptive technologies (Christensen 1997).

Centralised organisational structures are argued to restrict external inputs to creativity and innovation and encourage the "not-invented-here" syndrome. Decentralised structures are argued to ignore the longer-term investments necessary for major breakthroughs. Firms are experimenting with new forms of project-based organisation that help overcome the disadvantages for creativity and innovation of centralised and decentralised structures (Davies and Hobday 2005). These temporary coalitions of elements of numbers of firms disperse after the completion of the project, raising interesting questions about the ownership of any intellectual property created and how organisational memory is retained.

There are, therefore, a wide range of germane issues for management in the study of the relationship between creativity and innovation. Addressing these issues requires an appreciation of the contexts in which they

contribute; their economic imperatives, strategic directions and organisational structures; and the psychological and behavioural factors that
stimulate individual and team performance. It requires a combination of
psychological and organisational behaviour approaches to creativity and
the economics and strategy approaches to innovation.

THE DISCONNECT BETWEEN THE MANAGEMENT OF CREATIVITY AND INNOVATION IN THE LITERATURE

Study of the academic literature on the relationship between creativity
and innovation management shows that there is interest in the subjects in
a broad range of journals, that these journals tend not to be the highest
regarded academically and, despite the value in the connection between the
two, there is a profound disconnect between that which focuses on creativity and that on innovation.

Using Leximancer software, a content analysis was undertaken of the
three hundred articles in the journal *Creativity and Innovation Management* published between 1997 and 2007.[1] Of these, seventeen had the
words "creativity" and "innovation" in the abstract, and only eight had the
phrase "creativity and innovation", indicating that the two are treated quite
separately by a journal established expressly for their joint consideration.
A search of ABI Inform was conducted of over 2,840 social science journals excluding *Creativity and Innovation Management*: ninety-six journals
with 133 articles were found to include the phrase "creativity and innovation" in abstracts in the period 1997 to 2007; 308 articles had the words
"creativity" and "innovation" in their abstract. One journal had seven
articles with the phrase "creativity and innovation" in abstracts during this
period, one had four, seven had three, twelve had two and the remainder
had one article.

University of Queensland Business School (UQBS) operates a ranking
system for business and management journals allied to an incentive system. There is a close correlation between this and other business school's
rankings. Table 11.1 shows that more than half of the journals in which
creativity and innovation are included in the abstract are not ranked, i.e.
publication in them by UQBS staff would receive no institutional recognition. Eleven articles were included in the top two tiers. These journals
would be widely considered to be the leaders in the business and management field.

Leximancer maps the clustering of associated concepts in texts and Figure 11.1 shows the results of such analysis for three hundred full papers
in *Creativity and Innovation Management* published between 1997 and
2007. It can be seen that "creativity" and "innovation" are distantly related
and therefore the semantic relationships between the two concepts is limited within the papers. Reading of this visual representation of the text

Table 11.1   Ranking of Journals with "Creativity and Innovation" in Abstract

UQBS Ranking (1 high, 5 low)	Number of journals with articles
1	5
2	6
3	17
4	14
5	5
Unranked	49
Total	96

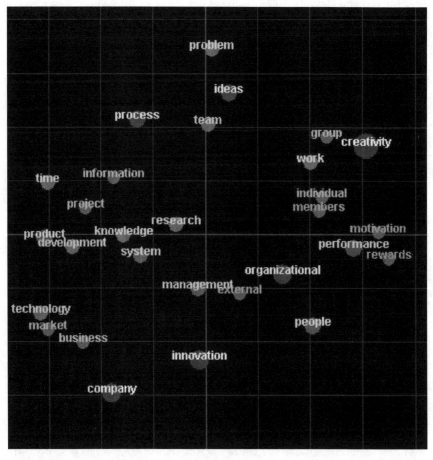

Figure 11.1   Results from Leximancer analysis.

from the journal demonstrates that innovation is couched in language representative of the organisation as a whole. Hence, its close relationship to the concepts "company", "employees", "research", "market", "business" and "technology". Creativity is related more closely to concepts that exist at the micro level of the team or the individual worker. These are concepts such as "motivation", "reward", "performance", "work" and "group".

To further confirm this disconnect, albeit in a much less empirically robust manner, the indexes of eight best-selling *Management of Innovation* texts were examined for references to creativity. Five had no references to creativity, two had one and one book had twelve references.

There is therefore compelling evidence to show that the literature on creativity and innovation management is disconnected. Given the range of potentially valuable connections described earlier, there are likely to be many possible ways of building bridges between the two areas of interest. We now turn to one particular means of establishing better connectivity. The following section considers the role of play around new technologies as a means of assisting the integration of creativity and innovation in the workplace and helping address some of the management issues identified earlier.

PLAY AS AN INTEGRATING CONCEPT

For some, play and work are highly distinct. Jean-Paul Sartre argues in *La Nausée* that meaningfulness is found in creativity and play and meaninglessness found in the order and discipline of work. For others, play and work are closely bound. Charles Dickens contended, "There can be no effective and satisfactory work without play; there can be no sound and wholesome thought without play" (quoted in Fielding 1960). The writer, Philip Pullman, says:

> It's when we do this foolish, time-consuming, romantic, quixotic, childlike, thing called play that we are most practical, most useful, and most firmly grounded in reality, because the world itself is the most unlikely of places, and it works in the oddest of ways, and we won't make any sense of it by doing what everybody else has done before us. It's when we fool about with the stuff the world is made of that we make the most valuable discoveries, we create the most lasting beauty, we discover the most profound truths. The youngest children can do it, and the greatest artists, the greatest scientists do it all the time (unpublished comment).

The latter perspective, and the notion that play at work is important for individual and organisational performance, is receiving increasing attention in the management literature (Schrage 2000; Dodgson, Gann and Salter 2005). This literature is emergent and shares none of the richness of the examination of play found, for example, in the fields of educational psychology, anthropology or cultural history. The management research challenge in the work context is to integrate what we know about *Homo Cogito* (man the thinker) and *Homo Faber* (man the maker), with an understanding of *Homo Ludens* (man the player). To do so remains a significant challenge. Play theorists have pointed out that the question "what is play?" is remarkably hard to answer: we all recognise play when we see it or engage in it, but it's a domain of life that is notoriously hard to conceptualise (Sutton-Smith 1997).

To begin to understand play, one can turn to Johan Huizinga's 1944 book *Homo Ludens: A Study of the Play Element in Culture.* In this delightfully insightful, contradictory and cantankerous book, Huizinga writes broadly about the relationship between play and language, law, war, knowing, poetry, philosophy and the arts. Play, he argues, is at once beautiful, fun and captivating and at the same time rule-bound, order-creating and tension-ridden. Play "creates order, *is* order. Into an imperfect world and into the confusion of life it brings a temporary, a limited perfection" (Huizinga 1970, 29). It occurs outside of real life and is essentially freedom-based: "all play is a voluntary activity. Play to order is no longer play; it could at best be but a forcible imitation of it" (1970, 26). Tension arises due to uncertainty and chanciness and the striving to decide issues.

Huizinga's contradictions come to the fore when considering the role of play at work. On the one hand, "it is an activity connected with no material interest, and no profit can be gained by it" (1970, 32). On the other, Huizinga notes the competitive sources and purposes of play that "appears when trade begins to create fields of activity within which each must try to surpass and outwit his neighbour" and how "some of the great business concerns deliberately instill the play-spirit into their workers so as to step up production" (1970, 226–27).

The relevance of Huizinga's historical and cultural analysis for management today is easily contested, although it would be diverting to consider the contemporary significance of playful demonstrations of virtue by "bragging and scoffing matches" (1970, 86) and "formal contests in invective and vituperation" (1970, 87). More relevant, perhaps, is his understanding of what today we would refer to as play in the virtual world. Play is "a free activity standing quite consciously outside 'ordinary' life executed within certain fixed limits of time and place, according to rules freely accepted but absolutely binding" (1970, 32). It "is not 'ordinary' or 'real' life. It is rather a stepping out of 'real' life into a temporal sphere of activity with a disposition of its own" (1970, 26). Joint participation in play brings "the feeling of being 'apart together' in an exceptional situation, of sharing something important" (1970, 31). Play, therefore, allows sanctioned free time for thinking and imagining and the exploration of new connections, but within boundaries broadly established by managerial objectives.

It is revealing that jazz improvisation is one of the most common recent metaphors used in the organisation and management literature to reflect this appreciation of the nature of play (see the special edition of *Organisation Science* on "Jazz Improvisation and Organizing" [1998, 9, 5]). Jazz improvisation provides an idiom for understanding the balance in the relationship between individuals as they collectively explore the unexpected within the confines of accepted styles and structures. As Weick (1998) points out, jazz reflects the way that effective improvisation, seen as spontaneous experiment, actually reflects depth of experience and degrees of discipline by its players. This musical idiom is used very practically by IBM in its massive annual "InnovationJam", designed to seek innovative ideas from tens of thousands of its staff, their families and company stakeholders (Dodgson, Gann and Salter 2008).

These aspects of play have enormous resonance with individual and collective computer-based games and is more widely assisted and enabled in the workplace by a new form of technology that helps place some of the processes of creativity, innovation and design in the virtual world.

Before exploring the virtual world it is necessary to state what the subsequent discussion is not going to address. It will not consider:

- the tension between individual and institutional interests in play at work
- the developmental characteristics of play in the sense of Piaget's child psychology
- psychological examination of the importance of play for happiness, identity and self-actualisation at work
- the role of playful games, jocularity and collegial badinage in improving social identification with workplaces
- the contribution of role-play, or assuming the personas of others to stimulate learning
- work–home relationships and the notion that as work becomes more playful, non-work, i.e. home life, becomes less playful and more work-intensive
- cultural and gender issues of play, and ownership of and relations to its artefacts (e.g. toys)

These and no doubt many other worthy research interests highlight the munificence of the subject for future study.

The focus of study here is to examine with a management perspective how play intermediates creativity and innovation to create value and competitive advantage. Play encompasses a number of different processes and mechanisms and includes those activities where people explore, template, model, prototype, rehearse and tinker with new ideas, often in combination with others with different skills in stimulating environments where work rules are relaxed.

DESIGN AND VIRTUAL REALITY TECHNOLOGIES

A number of new technologies, such as virtual reality, are helping creativity and innovation by facilitating play. This helps the search for optimal answers to technical and business problems and assists the convergence of the expectations of diverse stakeholders. Over recent years this kind of technology has significantly expanded its capabilities, aided by more powerful processors and sophisticated software, providing greater visualisation capacities. As this technology supports the innovation process, it has been described as a type of Innovation Technology (IvT; Dodgson, Gann and Salter 2005). The technology is routinely used both in the design of complex new products, such as the Boeing 787 and Airbus A380, and the relatively simple design of toothbrushes and shoes. Models and prototypes have always provided a mechanism for learning about artefacts before and after they have been built (Petroski 1985). They enable designers and engineers to examine different options and weigh the choices of form and function against one another (Thomke 2003). However, physical models and prototypes are expensive and time-consuming to create and are often unreliable, and digital tools are cheap and easily available (Schrage 2000; Thomke 2003). Computer-based digital models and virtual prototypes assist in abstracting physical phenomena, allowing experimentation, simulation and play with different options whilst engaging with a range of interested parties.

Innovative projects are often complex, multidisciplinary and collaborative. IvT can help the management of projects by providing the means by which various forms of integration can occur between components of systems, different skills and knowledge bases and organisations. If, as Simon (1996) suggests, complex problems are solved by breaking them down into smaller components, IvT can assist in both analysis at the decomposed level and in the reintegration of those components. It helps ensure technical compatibility and provides cost-effective ways of manipulating and playing with creative potential solutions to problems.

A core element of IvT is its capacity to represent information visually. The capacity to visualise something in three and four dimensions is a much better aid to comprehension than information presented in papers and diagrams. As the previous head of technology strategy at IBM says, "Visualisation is the broadband into our brains". By way of example, contrast the client of an architect confronted by technical line drawings of her home with one provided with a virtual reality experience. Architects regularly use "walkthrough" representations of buildings with clients prior to their construction, allowing their customers to get a feel of the layout and ambience of their homes or offices before a brick is laid.

The growth in commercial use of the Internet has created a massive new set of opportunities for new technologies to support creativity and innovation. In its early days, Web 1.0 technologies provided some useful capabilities in the search for data, but did not enable the social and business interactions that are all-important for creativity and innovation. New Web 2.0 technologies such as Second Life and social networking sites are being introduced which have provided a new set of opportunities for on-line communities of innovators (Jeppesen and Frederiksen 2006; Dahlander and Wallin 2006).

The use of a range of Web-based systems—including blogging, wikis, online encyclopaedias and podcasting—is creating an environment in which people develop their ideas in a more horizontal, collaborative way than before. Some of these tools are being developed following success in the market for multiplayer games involving huge numbers of players. Other Web 2.0 technologies such as "mash-up" software are being developed to enable innovators to capture, combine and analyse data from different sources online. They may provide new ways for firms to better relate to markets and understand highly engaged and active customers expecting to be involved in the development of the products and services they want. This high level of engagement of users in the creation of the innovations they desire has led to its description as the "democratisation" of innovation (von Hippel 2005).

In the future, the possibilities envisaged for Web 3.0 technologies, including immersive systems that enable firms to share "mixed-reality" (concurrent real and virtual) experiences of possible products and services will improve the way choices can be honed through interaction with different user reference groups. These simulated and immersive systems will involve suppliers and partners modelling value maximisation amongst various options and the equitable distribution of their results (Gann and Dodgson 2007).

Let us turn to some case study examples of play at work assisting to merge creativity and innovation with particular reference to the ways new technologies are assisting the process. The following vignettes resulted from the work of the author with two colleagues, David Gann and Ammon Salter, and the full richness of the cases is published elsewhere.

The Playful Design Company[2]

IDEO is a highly successful provider of design and innovation services, employing about five hundred people in 2006, in offices around the world. It has built a reputation for helping other firms to innovate in their products and services by applying creative techniques learnt in the design studio and design school environments. The company combines "human factors" and aesthetic design with product engineering knowledge to produce products for firms from Apple to Nike to Prada. Its designs include the computer mouse, Palm V, and a range of cameras and toothbrushes. It designed the whale that starred in the film *Free Willy*. IDEO has contributed to the design of over three thousand products and works on sixty to eighty products at any one time. IDEO has been described by *Fast Company* magazine as the "world's most celebrated design firm"; by the *Wall Street Journal* as "imagination's playground"; and *Fortune* described its visit to IDEO as "a Day at Innovation U".

To enable the company to deal with so many diverse projects, it recruits a wide range of talent, and also enjoys special links with the Stanford University Institute of Design. It employs graduates from psychology, anthropology and biomechanics as well as design engineering.

The leaders of IDEO have a very high profile in the international design community. They claim to have a creative culture—"low on hierarchy, big on communications and requiring a minimum of ego"—that uses a "collaborative methodology that simultaneously examines user desirability, technical feasibility and business viability, and employs a range of techniques to visualise, evaluate and refine opportunities for design and development, such as observation, brainstorming, rapid prototyping and implementation". IDEO sells its design methodologies to other companies in the form of courses and training materials. It possesses a large repository—a "toy box"—of devices and designs from a wide range of products which staff play with when seeking solutions to new problems. It is particularly skilled at playing with creative ideas developed for one industry or project to explore their innovative application in others. Play in this environment enables the cross-fertilisation and the serendipitous connection of unrelated ideas.

The Architectural Practice[3]

Frank Gehry's architectural practice has been responsible for a number of landmark buildings, including the Bilbao Guggenheim and Walt Disney Concert Hall. It was one of the first architectural firms to adopt computer-aided design (CAD) to integrate its design processes and link these to virtual and rapid prototyping systems. It has an associated company—Gehry Technologies—which has collaborated with other firms to create a software tool called Virtual Building.

Virtual Building provides a range of modelling and simulation techniques, including city planning, simulation of transportation and movement of people in the urban environment, as well as more traditional engineering processes such as thermal, daylight and acoustic simulation, using computer models of physical properties. This assists the visualisation of problems and solutions in design, construction and operations using computer simulation. It provides what has become known as a "single digital environment" in which data from different disciplines are added once, and then integrated in a way that others can see and use them. The digital environment encompasses design, engineering, analysis, fabrication, project management and on-site construction activities. These are represented in the virtual, digital model. The model enables more effective communication between different members of project teams, helping to foster a collaborative working environment in which new ideas for innovation can emerge through interdisciplinary endeavour. These ideas can be shared and developed by groups of specialists who previously would have had no means of connecting with one another as a team. This form of technology also helps to capture ideas as they are being developed. With the right organisational structures and skills, these technologies can enable better choices of solutions to be made, capturing and sharing lessons and optimising design decisions on applicability and cost.

Frank Gehry has said that creativity is about play and he commented upon the combination of new technology and processes that have enabled him to produce his innovative, malleable, plastic form of building design by saying that:

> This technology provides a way for me to get closer to the craft. In the past, there were many layers between my rough sketch and the final building, and the feeling of the design could get lost before it reached the craftsman. It feels like I've been speaking a foreign language and now, all of a sudden, the craftsman understands me. In this case, the computer is not dehumanising, it's an interpreter.

Play in this case supports the creative imagination and allows specialists from different backgrounds with different skills to combine to deliver innovative designs.

The New Product Development Group[4]

CreateInnovate is a small group within Procter and Gamble (P&G) producing new packaging. Packaging design is hugely important in consumer product markets: it helps sell products. CreateInnovate uses technology to increase innovation in the packaging and marketing of P&G's products through the combination of different sources of knowledge within and outside of the firm in the process of design, through the use of a visualisation suite to test representations of package designs with consumers, and through a 3D CAD system which can simulate and model prototypes, linked to rapid prototyping technologies.

The group spends a lot of effort brainstorming internally and externally with design houses such as IDEO. The aims of these brainstorming exercises are to bring together a number of multiskilled people with different but relevant knowledge and experience, in a structured way, in order to create ideas. Ideas for new packaging concepts are described in text and in sketches, often with the help of sketch artists. From these rough sketches,

3D CAD models are quickly produced. This CAD model will be used throughout the entire development and manufacturing process: from initial concept to Computer Numerical Control manufacturing. As the head of the group says, "this digital model will live with the product idea until it goes to manufacturing. We only create one model and then play with it."

Once the model is produced and virtual product testing begins, teams of people from around the world examine the virtual model and comment on what they like and what they do not like. The image is very close to the real thing. As the head of the group puts it:

> Once it has been created digitally you can show it to people and see what they think. You can change it and tweak it to see if people's attitudes towards it improve. Using the new technology helps us get a richer representation of the new idea and almost immediate feedback from users. The whole process takes days not months. The new tools allow people to play easier. In some ways, the virtual prototypes are like games where users can play with the design, exploring different design options. It provides vivid, accurate representation of the project. It allows you take one hundred ideas and kill the ninety-eight of them that are no good.

The use of simulation tools has changed the staffing profile of the group. It previously used almost exclusively to hire people with a relatively narrow background in chemistry, biochemistry or chemical engineering. As it become involved more and more in design, it started to hire a broader church of people. The group has, for example, hired industrial designers and mechanical engineers. The capacity of technology to represent ideas and prototypes quickly and cheaply assists the interdependencies and knowledge flows between this diverse range of people.

This case illustrates the value of playing with creative techniques for visualising and evaluating innovations and enabling "creators" to get close to engineers and manufacturers of final products.

The New Engineering Field[5]

Since 9/11 significant attention has been placed on designing buildings to withstand extreme events. New knowledge and practices have been developed about fire engineering to help improve the safety of buildings. Knowledge of building design and the behaviour of people in emergencies has expanded such that underlying design concepts of fire safety have changed significantly. Innovations, including the use of elevators to evacuate people, have been developed with the aim of avoiding loss of life associated with catastrophic events such as the collapse of the World Trade Center. As an emerging field of practice, fire engineering builds upon the integration of diverse skills and knowledge of architects, design and construction engineers, building clients, regulators, fire authorities and fire-fighters. These multiple players adopt different and evolving positions on the development of the field. Fire engineering is a state of practice described as being "in ferment", where the field is seeking credibility in the face of the unknown. The providers of fire engineering services, for example, are obviously active proponents of the field whilst regulators and fire-fighters are understandably concerned about any innovations with implications for safety. Regulators take a lot of persuasion to change established practice. Yet practices are changing and fire engineering is becoming a better-established field.

In some cases, visual conversations and play allow fire engineers to gain regulatory approval for a new design. The visual power of the simulation technology used, for example, helps transform regulators' understanding of new practices, overcoming their tendency towards "precedent" and "repetition" in design and support the realisation of innovative solutions. Play around the technology contributed to the changing nature of the conversation between regulators and designers and the shifting boundaries of what is acceptable. The case illustrates how play can help coalesce creative ideas amongst disparate groups to produce radically innovative approaches.

DISCUSSION

Play is an activity that can connect creativity and innovation and is receiving more attention to its capacity to do so in the organisation studies literature (Mainemelis and Ronson 2006). There is value in studying this relationship from a management perspective, theoretically and practically.

Theories of creativity and innovation tend to derive from very different bodies of literature: primarily psychology and organisational behaviour in the former and economics and industrial organisation in the latter. Analysis of play can draw on both groups of theories by connecting explanations of individual and team actions with corporate objectives.

Practically, play can generate ideas and produce options to be tested for subsequent execution: it is the intermediary between thinking and doing. An organisational commitment to play, seen in the provision of time for exploration and the creation of an environment where mistakes and failures are tolerated and relatively inconsequential (i.e. fast and cheap), facilitates innovation. The short, descriptive case studies—explored in much greater detail elsewhere—illustrate the role that new technologies can assume in bridging the creation of ideas and their application. The technologies discussed are described by Dodgson, Gann and Salter (2005) as "innovation technologies" because of their capacities to help enable the integration of all the various contributions to the innovation process. These were used to connect and combine creativity and innovation in highly innovative projects and products and by individuals, teams, firms and groups of collaborating organisations.

The use of the technologies resonated with the playing of "games" with work tasks, contributing to the fun of creativity and innovation. In these business environments their use was associated with the usual competitive tensions and, in the case of developing safety-critical fire engineering, significant pressures. They helped contribute to the imposition of a certain discipline and order through formal design or new product development processes. Although its use in the virtual world occurred in a form unimaginable to Huizinga, the virtuality of the play he described is important as it steps out or the "real" and "ordinary" within the constraints of binding rules. It allows for experimentation and imagination within a self-imposed ordered environment. In Bilton's (2007) terms, it exerts managerial order into the chaos of creativity. When it comes to creativity and innovation, playing in the virtual world is cheaper and less potentially harmful than playing in the real world. Play creates boundaries and constraints on creative processes and focuses them on the organisational objectives of delivering innovation.

Playing with IvT assists with incremental and radical innovation, and by facilitating the integration of design and manufacturing through common digital platforms helps link the exploration and exploitation efforts of organisations. It provides an "ambidextrous" means of integrating those organisational structures tasked with creating new options with those responsible for their implementation.

In the virtual world of play, group learning is facilitated by bringing together people from different disciplines and professional backgrounds with different skills. We saw how the skills profiles of firms using this technology are diversifying. Ideas are transferred—and connections and combinations occurred—between individuals with various ways of thinking, and from and between diverse groups: universities and customers and multiple parties. This helps the search for creative solutions and prevents introspection within teams and with innovation ecosystems of coexisting and cooperating firms.

Technology is simply an artefact, useless without imagination, skills, organisation and purposeful application. It may facilitate and enable play, but its most creative use will lie in the exploration of individuals and groups unbounded and unfettered by restrictive organisational discipline. Order derives more from the rules in the game—in this case the capacities of the technology—rather than decisions on how the game is to be played. Although IBM has recently developed its Virtual World Guidelines, advising staff how to conduct themselves when they are interacting virtually, it remains keen not to restrict the excitement it engenders and "does not want to act like a sheriff" (*USA Today*, 26 July 2007). Playful activities will more successfully link creativity and innovation when they are fun.

Whilst having more fun might not strictly comply with the direct objectives of public policy (although there is considerable interest in measuring happiness), its relationship with innovation and productivity should ensure policy interest. This applies within the public service itself, of course, where experimentation and tolerance of failure with public money is rarely endorsed, although the benefits of both may be conducive to more imaginative and effective policy-making. It has implications for educational policy, where as we have seen, the capacity to share expertise across organisational boundaries is increasingly important and inter- and multidisciplinary skills are critical. It has implications for innovation policies designed to encourage innovation in the private and pubic sectors and in their various forms of collaboration. By facilitating communication between disparate groups, and enabling greater, quicker, cheaper and more adventurous experimentation, it stimulates and intensifies the innovation process.

ACKNOWLEDGMENT

The author would like to acknowledge the insights and comments on this chapter from Catelijne Coopmans, David Gann, Jane Marceau, Leon Mann and Jennifer Whyte. Thanks are extended to Stuart Middleton for his comments and data analysis. Janet Chan's insightful and incisive criticisms deserved better responses.

NOTES

1. This journal states its objective is to fill "a crucial gap in management literature between the theory and practice of organisation imagination and innovation".
2. This section is taken from Dodgson, Gann and Salter (2008).
3. This section is taken from Dodgson, Gann and Salter (2008) and Dodgson, Gann and Salter (2007b).
4. This section is taken from Dodgson, Gann and Salter (2006).
5. This section is taken from Dodgson, Gann and Salter (2007a).

BIBLIOGRAPHY

Amabile, T. 1996. *Creativity in context*. Boulder, CO: Westview Press.
———. 1998. How to kill creativity. *Harvard Business Review* 76 (5): 76–87.
Belbin, R. M. 1993. *Team roles at work*. Oxford: Butterworth-Heinemann.
Bilton, C. 2007. *Management and creativity: From creative industries to creative management*. Malden, MA: Blackwell Pub.
Boden, M. A. 1990. *The creative mind: Myths and mechanism*. London: Routledge.
Braverman, H. 1974. *Labor and monopoly capital: The degradation of work in the twentieth century*. New York: Monthly Review Press.

Breschi, S., and F. Malerba. 2005. *Clusters, networks, and innovation*. Oxford: Oxford University Press.

Burns, T., and G. Stalker. 1961. *The management of innovation*. London: Tavistock Publications.

Caves, R. 2000. *Creative industries: contracts between art and commerce*. Cambridge, MA: Harvard University Press.

Chee, L. S., and T. H. Phuong. 1996. Managing diverse groups for creativity. *Creativity and Innovation Management* 5 (2): 93–96.

Chesbrough, H. 2003. *Open innovation: The new imperative for creating and profiting from technology*. Cambridge, MA: Harvard Business School Press.

Christensen, C. 1997. *The innovator's dilemma: When new technologies cause great firms to fail*. Boston, MA: Harvard Business School Press.

Cooke, P., and K. Morgan. 2000. *The associational economy: Firms, regions, and innovation*. Oxford: Oxford University Press.

Dahlander, L., and M. W. Wallin. 2006. A man on the inside: Unlocking communities as complementary assets. *Research Policy* 36 (6): 1243–59.

Davies, A., and M. Hobday. 2005. *The business of projects: Managing innovation in complex products and systems*. Cambridge: Cambridge University Press.

DiMaggio, P., and W. Powell. 1983. The iron cage revisited: Institutional isomorphism and collective rationality in organisational fields. *American Sociological Review* 48: 147–60.

Dodgson, M. 1993. *Technological collaboration in industry: Strategy policy and internationalization in innovation*. London: Routledge.

Dodgson, M., D. Gann and A. Salter. 2005. *Think, play, do: Technology, innovation and organisation*. Oxford: Oxford University Press.

———. 2006. The role of technology in the shift towards open innovation: The case of Procter and Gamble. *R&D Management* 36 (3): 333–46.

———. 2007a. "In case of fire, please use the elevator": Simulation technology and organization in fire engineering. *Organization Science* 18 (5): 849–64.

———. 2007b. The impact of modelling and simulation technology on engineering problem solving. *Technology Analysis & Strategic Management* 19 (4): 471–90.

———. 2008. *The management of technological innovation: Strategy and practice*. Oxford: Oxford University Press.

Fielding, K. 1960. *The speeches of Charles Dickens*. Oxford: Clarendon Press.

Florida, R. 2002. *The rise of the creative class*. New York: Basic Books.

Gann, D., and M. Dodgson. 2007. *Innovation technology: How new technologies are changing the way we innovate*. London: National Endowment for Science, Technology and the Arts.

Helfat, C. E., S. Finkelstein, W. Mitchell, M. Peteraf, H. Singh, D. Teece, and S. Winter. 2007. *Dynamic capabilities: Understanding strategic change in organizations*. Malden, MA: Blackwell.

Huizinga, J. 1970. *Homo Ludens: A study of the play element in culture*. New York: J&J Harper.

Jeppesen, L., and L. Frederiksen. 2006. Why do users contribute to firm-hosted user communities? The case of computer-controlled music instruments. *Organization Science* 17 (1): 45–63.

Kim, L. 1997. *Imitation to innovation: The dynamics of Korea's technological learning*. Boston: Harvard Business School Press.

Mainemelis, C., and S. Ronson. 2006. Ideas are born in fields of play: Towards a theory of play and creativity in organizational settings. *Research in Organizational Behavior: An Annual Series of Analytical Essays and Critical Reviews* 27: 81–131.

March, J. 1991. Exploration and exploitation in organizational learning. *Organization Science* 2 (1): 71–87.

Marx, K. 1976. *Capital,* Volume 1. Harmondsworth: Pelican.

McFadzean, E. S. 1998. Enhancing creative thinking within organisations. *Management Decision* 36 (5): 309–15.

Petroski, H. 1985. *To engineer is human: The role of failure in successful design.* London: Macmillan.

Potts, J. 2009. Creative industries and innovation policy. *Innovation: Management, Policy and Practice* 11 (2): 138–47.

Puccio, G. 1999. Creative problem solving preferences: Their identification and implications. *Creativity and Innovation Management* 8 (3): 171–78.

Schrage, M. 2000. *Serious play: How the world's best companies simulate to innovate.* Boston: Harvard Business School Press.

Schumpeter, J. 1942/1975. Capitalism, socialism and democracy. New York. Harper

Sigurdson, J. 2005. *Technological superpower china.* Cheltenham: Edward Elgar.

Simon, H. A. 1996. *The sciences of the artificial.* Boston: MIT Press.

Snow, T. A., and J. D. Couger. 1996. Process: Creativity improvement in a system development work unit. *Creativity and Innovation Management* 5 (4): 234–40.

Sutton, R. 2006. *The no asshole rule: Building a civilized workplace and surviving one that isn't.* New York: Warner Books.

Sutton-Smith, B. 1997. *The ambiguity of play.* Cambridge, MA: Harvard University Press.

Thomas, K. W. 1983. "Conflict and conflict management". In *Handbook of industrial and organizational psychology,* ed. M. D. Dunnette, 889–935. New York: Wiley.

Thomke, S. 2003. *Experimentation matters.* Boston: Harvard Business School Press.

von Hippel, E. 2005. *Democratizing innovation.* Cambridge, MA: MIT Press.

Weick, K. 1998. Improvisation as a mindset for organisational analysis. *Organisation Science* 9 (5): 543–55.

Woodward, J. 1965. *Industrial organization: Theory and practice.* London: Oxford University Press.

12 Inducing and Disciplining Creativity in Organisations under Escalating Complexity

Jonathan West

The origins of organisational barriers to creativity and innovation have long been a subject of research in organisational studies. I argue here that organisations struggle to perform creative and innovative tasks because of unintended consequences of organisational systems introduced to deal with information, which over time increases in complexity and uncertainty. The problem is that the goal for organisations in respect of innovation is simultaneously to encourage creativity while at the same time integrating that creativity into the organisation's operations. I argue that this dual challenge places contradictory tensions and demands upon organisations, which increase with complexity and uncertainty. The range of organisational procedures and approaches to "manage" creativity and the flood of management fads to "foster creativity" can be understood as a response to these contradictions.

In this chapter I explore the terms of these contradictory demands, through: (a) consideration of the character of the information with which organisations must cope as they attempt to innovate, (b) discussion of the two generic organisational structural responses to that information and (c) an analysis of the structure of the problem-solving process, itself a form of information-processing, which is at the heart of innovation by organisations.

My main claim is that the two principal organisational responses to complexity and uncertainty—toward centralisation on the one hand and decentralisation on the other—both tend to dampen creativity and impede innovation, but in different ways. These organisational systems for coping with complexity and uncertainty are poorly adapted to induce or sustain creativity, and can retard it. In addition, complexity and uncertainty have increased over time, bringing an increase in dysfunctional organisational responses that interfere with the information-processing required for creative activity.

INNOVATION AND CREATIVITY DEFINED

I use the term "innovation" to mean the introduction to actual use of new products, services or production processes, that is, a change to an economic process. The essential challenge in bringing new products or services to use is to overcome the wide diversity of obstacles encountered along the path, that is, of posing and solving the "right" problems. "Creativity" is a vital aspect of problem (or "opportunity") identification and solution because it is the source of novel approaches. Creativity can thus be viewed as a sub-category of problem-solving, in which novel options or novel methodologies are introduced and evaluated. Creativity is not the production of new information alone, but the recognition and novel interpretation of this information.

A PROBLEM-SOLVING PERSPECTIVE ON CREATIVITY AND INNOVATION

The capability to both raise and overcome problems is at the heart of successful innovation, and understanding the sources of difficulty in problem selection and solution is our window into the nature of organisational inducement and integration of creativity. A problem-solving perspective (understood to include the identification and appropriate selection of problems) offers several advantages to the investigator of creativity and innovation.

First, this perspective directs attention towards the key tasks an innovation system must perform, and helps us understand the dynamic nature of the innovation process (see West, Chapter 2, and Marceau, Chapters 3 and 4, this volume). By understanding the evolving range of problems that economic institutions and organisations must confront, along with the structure of these problems, we can gain insight into the nature of the capabilities required to select "good" problems for solution, the creative capabilities required to resolve them and those needed to implement the chosen solutions (Dosi and Marengo 1993).

Second, because the problems inherent in innovation efforts can be many and varied—and I will argue that the evolution of technology is increasing both the quantity and variety of problems—a widening range of expertise and types of creativity are required to innovate. Coordination of this range of expertise calls for increasingly sophisticated organisational and institutional arrangements. A problem-solving perspective thus enables analysis of innovation as the outcome of social and technical *systems*, and as a set of creative and problem-solving capabilities possessed by people, organisations and institutions, as well as the interactions among those capabilities. It thus assists us to see innovation as a particular type of social process, rather than simply as individual change or as an embodied outcome.

Third, a problem-solving view frees us from an excessively technical view of the innovation process. It directs attention to the social, organisational and cognitive issues inherent in innovation, as creativity within these fields is combined to resolve the multiplicity of challenges presented. This variety of capability makes up the elements of an effective innovation system, each of which must be recognised, analysed and evaluated if the innovation process is to be understood.

Fourth, a major advantage offered by a problem-solving perspective is that it assists an understanding of innovation in non-material systems, such as the service industries that comprise the larger part of modern economies. To understand innovation in the service sector, we must view the task not only as physical transformation—the processing of things—but as the transformation of ideas and information. Information-processing systems exhibit characteristic attributes different from those of physical-processing systems. Indeed, in contemporary economies even traditional manufacturing is information-based; very few workers today directly transform a physical object. Most work is intellectual: designing products or processes, guiding and supervising machines, posing and solving problems.[1]

These distinctions are important in understanding the nature of organisational and problem-solving tasks characteristic of physical and non-physical systems. Manufacturing-based businesses tend to be physical-asset-intensive; exhibit economies of scale and scope; be dominated by "replication", with replication's attendant steep learning curves; and employ many workers of diverse educational backgrounds, organised through large interconnected systems. Information-based businesses tend to have high revenue-to-fixed-asset ratios; exhibit network economies; be "reconception" rather than replication oriented; and employ fewer employees, with bimodal educational backgrounds (split between university-educated professionals and low-skilled staff), organised through small groups and projects (Austin, Devin and West 2000).

Finally, a problem-solving framework points to a means to evaluate the performance of innovation systems. By proceeding from the nature of the tasks to be performed, a problem-solving perspective allows us to assess the health of an innovation system in terms of its ability to meet the challenge of the responsibilities it must perform.

COMPLEXITY AND UNCERTAINTY IN ORGANISATIONAL INFORMATION-PROCESSING SYSTEMS

Information-processing is difficult to the extent that tasks are characterised by complexity and uncertainty (Arrow 1974). In the absence of complexity and uncertainty, information-processing can be made routine and even automated (Winter 1987). Complexity here can be defined as the number of variables, and variable interactions, with which an organisation must contend (Simon 1962). Uncertainty can be defined as the "perceived inability to predict something accurately because of a lack of information or an inability to discriminate between relevant and irrelevant data" (Milliken 1987, 136). Technological novelty often induces uncertainty.

More Complexity Produces More Uncertainty

Greater complexity itself can also create greater uncertainty. As the number of variables that must be processed increases, and along with them the number of interactions, the range of possible outcomes from a specific project mounts. In addition, as the breadth of knowledge expands, so too does the real uncertainty faced by aspiring innovators. Uncertainty increases in severity from the question, "what state will certain (known) variables assume in the future?" through "which variables are important?" to the most severe, "what information is needed to determine which variables will be important?"

Increased uncertainty produces greater demands on creativity. But most organisational systems designed to cope with complexity and uncertainty focus on routinising operations. And, as we shall see, growth magnifies this difficulty.

The issues in organisational information-processing are more than technical; they are also organisational. As technical complexity and uncertainty increase, organisations undergo mounting information-processing load and stress. The stress stems from the demands of complex and uncertain technical issues and from the ensuing increase in organisational-structure complexity. As the number and range of technical variables to be controlled increases, and therefore sources of relevant information become diversified, the division of labour for successful task completion becomes more differentiated. A greater range of creative and scientific expertise becomes necessary; the work of highly specialised staff must then be allocated and the results must be synthesised and combined. In short, technical complexity induces organisational complexity, and organisational complexity creates the need for additional information-processing (Calvo and Wellisz 1978).

As complexity and uncertainty escalate, at some point existing organisational information-processing systems and structures begin to fail. Either they drown in the sheer volume of information the organisation must process, or their structures are overwhelmed by the challenge of organisational coordination. Symptoms of information-processing breakdown and coordination failure include confusion, delay and redundant effort. The improving performance of competitors and rivals may precipitate a crisis. Examples include military organisations overwhelmed in the "fog of war"—itself a dramatic term for complexity—and business organisations with hopelessly delayed projects due to the number of features. Examples include Microsoft with its new operating systems and Lockheed Martin with its F-35 Joint Strike Fighter aircraft.

Tendency toward Escalating Levels of Complexity and Uncertainty

The challenge of organising for creativity and innovation rises as complexity and uncertainty increase across a wide range of technologies and social challenges. From automobile components and aircraft components, to handguns, transistors in microchips, lines of code in software operating systems and PBX interconnects, the number of system elements necessary to provide functionality—or components included in a specific device—has grown over time, often exponentially. Some examples: the number of components in a typical automobile increased between 1920 and 2003 from fifteen hundred to thirty thousand; the number of transistors in an average computer microchip from one thousand to one hundred million between 1970 and 2003; and the number of components in a typical aircraft from twenty thousand to 3.5 million between 1945 and 2003.

Complexity and uncertainty increase to meet the rising demands of information-processing induced by technological advance. Information technology, including the advent of computers, can provide a partially offsetting force, but as the functionality of technology grows so too does the number of organisational-system elements required to deliver that functionality. As the range and depth of technological knowledge grows, so too does pressure on organisational members to specialise.

In short, technical specialisation induces human specialisation, and in turn organisational complexity. These elements in interaction can produce immense levels of system complexity. A system comprised of only fifteen elements, for example, will have fifteen linear interactions, but 120 possible non-linear dyadic interactions, and 65,535 potential non-linear combinations.[2] Technical specialisation accounts for a form of organisational complexity I call "horizontal complexity", the number of functional or discipline specialties an organisation must embrace. The need to integrate specialties generates another kind of complexity, "vertical complexity", the number of levels in the organisational hierarchy needed to coordinate specialties. The number of levels is determined by the feasible span of control over diverse specialties manageable by an organisational integrator. The more knowledge-intensive and novel the work, *ceterus paribus*, the smaller the feasible span of control (Keren and Levhari 1979). The smaller the feasible span of control, the larger the number of vertical layers required, and the greater the total organisational complexity.

Unfortunately, further complexity in the organisational system produces declining marginal returns in problem-solving capability. At first, additional complexity delivers greater benefits to the organisation than it costs. More finely differentiated divisions of labour enable the organisation to cope with more complex and difficult problems. But additional complexity comes at a price. The hierarchy must be resourced, and information decays as it moves through layers of decision-makers. Over time, the costs of additional organisational complexity begin to overwhelm the benefits. Ultimately, the organisation must devote a disproportionate amount of its resources to self-maintenance and begins to suffer symptoms of unmanageable complexity.[3]

Accordingly, modern innovation systems have become exceedingly complex, manifest a general tendency to escalate in complexity and pose mounting problems for managers. None of this will surprise, although the acceleration and pervasiveness of the phenomenon might surprise. A core issue in the economics of innovation is to understand the tools available to managers and policymakers to cope with the organisational problems created by complexity.

Two Organisational Responses to Complexity and Uncertainty

As complexity and uncertainty grow, organisations can choose between two alternative paths. They can either increase their information-processing capability or they can reconfigure their tasks to solve problems with less information. The first path usually prompts a centralised strategy; a single organisational unit takes on the information-processing role to make decisions on behalf of the entire organisation. The option of simply increasing the technical information-processing capacity of all organisational subunits tends to generate more information and intensifies coordination and integration challenges.

The second path leads to decentralisation, aimed to reduce the central demand for information and information-processing by distributing the task among semi-autonomous subunits. But this approach has implications for managing coordination among units, which itself involves information acquisition and information-processing. A means for coping with the extra information-processing load required for managing coordination between units is to resort to standard guidelines and principles such as the "design rules" engineers use to manage the interaction among elements in a technical system (Baldwin and Clark 2000).

These two paths—towards centralisation or decentralisation of information-processing—present different challenges to managers. It is difficult for organisations to take both simultaneously. The first—centralisation— implies tightening control of information, increasing its flow to a single location for processing, and communication of results to other units for action. It also implies design of the organisation's tasks so they can be performed by centralised structures. The second—decentralisation—calls for the wider distribution of information and the design of tasks so they can be performed semi-autonomously. Each solution generates a distinct set of organisational problems and risks, and also pressure on organisations to adapt their structure and culture accordingly.

And both organisational strategies—centralisation and decentralisation—tend to curtail creativity, but in different ways. Centralisation limits the range of personnel and organisational subunits "permitted" to define problems, collect information, and make decisions. Staff outside the central information-processing unit are required to implement decisions and abide by rules and procedures. Novelty and exploration become the province of a single centralised unit. The result is that creative people are often frustrated working in the sub-units. The problem worsens as complexity and uncertainty escalate, and stricter information-reporting and implementation procedures are enforced. Under decentralisation— because the tasks to be undertaken by different organisational sub-units must be prescribed in advance—the range of novel projects the organisation can undertake is circumscribed. In addition, as modes of coordination between units must be prescribed, so too are potential forms of information integration limited. In sum, both responses to dealing with complexity tend to discourage creativity, and both are exacerbated by greater complexity and uncertainty.

Organisations can choose from a wide variety of counter-measures to these challenges. Decentralisation proliferates the number and range of sites, creating, for example, laboratory or factory networks. This response is widely adopted, especially as organisations grow in overall size and output. It raises, in turn, several difficult managerial issues: which activities should be centralised and which decentralised; what should be the basis of grouping or focus; and how should the subunits relate to one another (for example, by competing or cooperating)?

A currently popular organisational form, the project-based organisation, can be seen as an attempt to overcome some of the problems of complexity. The organisation decentralises along project types. But often the problem of complexity is simply displaced to other levels in the organisation, and similar questions arise: What level of the organisation— who—should select the projects (top-down versus bottom-up)? How are disparate projects to be integrated and connected with one another? What organisational structure will ensure that tasks are not duplicated, or fall into a lacuna?

Each path thus generates characteristic and specific problems. Centralisation risks information overload at the centre and confusion among proliferating communication channels. Symptoms include bottlenecks in information flow, response delays and errors due to information-veracity decay. Decentralisation risks poor communication among sub-units, and overlapping or uncoordinated activity. Symptoms include confusion and conflict within the organisation, inefficiency and redundant or neglected task performance (Van Creveld 1985).

The resort to decentralisation often results from a search for simplicity. Simplicity under decentralisation implies *focus*, dedicating each unit to a particular form of task. Organisational focus can be attempted on many dimensions: *geographic* focus, for example, when complexity derives from transport logistics or regional preferences; *product* focus when complexity derives from component integration or customer-preference issues; *process* focus when complexity derives from production technology or characteristic learning curves; *life-cycle-stage* focus when complexity derives from product variety or ramp-up demands; and *volume* focus when complexity derives from scale differences at different phases of a product market's development (Skinner 1974). The dilemma for organisational creativity becomes most apparent when we examine the elements of typical organisational problem-solving processes. All are overshadowed by complexity and uncertainty.

THE STRUCTURE OF PROBLEM-SOLVING

Effective information acquisition and analysis is the essence of posing and solving problems. The information-processing perspective has been applied to economic activity for many years (Bolton and Dewatripont 1994). Any information-processing activity directed at solving problems is an effort to assemble, sort and transfer data or other forms of information, ranging from impressions, through concepts, to measurements (Nelson 1961). As such, information-processing systems must include mechanisms for: (a) identifying and defining appropriate problems; (b) selecting relevant options for inclusion in a set to be formally evaluated; (c) suitable means to test and eliminate sub-optimal options; and (d) means for transferring solutions for implementation (Newell and Simon 1972). Each step in the process depends upon creativity, especially the first.

Selecting the Right Problem

Choosing the right problem is the starting point of innovation. Too often prospective innovators waste time, resources and effort working on poorly conceived problems. From the firm's point of view, the "problem" to address may be little more than its own need to generate new customer offerings to increase profitability. Yet a common source of failure in firm-level innovation is the tendency to solve "problems" and produce products that don't matter to potential customers.

An important distinction should be made here between "problems" narrowly conceived (something going wrong) and "opportunities" (a potential new field of activity). The most powerful forms of innovation stem from identification of hitherto unrecognised opportunities that lead to creation of new markets or new business models. These forms of innovation are often the most value creating. The strongest innovators, therefore, are those best at conceiving the "right" problem and designing a novel approach to its resolution, rather than those with the greatest scientific or engineering expertise.

Inducing creativity is at the heart of this challenge. Recognising unexploited opportunities calls for creativity. Enhancing the capability to identify opportunities is a vital but difficult part of organisational innovation systems. Much of the challenge is the fact that problem identification is usually a social rather than individual process, and calls for the combination of many different types of expertise. Integrating that expertise is a vital skill possessed by successful innovators (West and Iansiti 1999).

It is here, at problem selection, that organisations attempting to cope with complexity and uncertainty first tend to adopt procedures that impede innovation. A common approach to managing organisational resource allocation is to subject all proposals for resource commitment to a single simple evaluative system—an investment screen, commonly in the form of a "hurdle rate". The problem with such procedures is that they usually subject both highly creative proposals and routine proposals to the same test. The effect is that creative projects can be seen as excessively risky and fail to gain management support. An alternative approach is to develop a balanced *portfolio* of potential innovation projects, some more risky than others, to be subjected to different types of scrutiny (Wheelwright and Clark 1992a).

Another key finding of research on problem identification is that innovation problems are often defined by customers rather than inventors. Indeed, customers commonly suggest both the problem and the solution (von Hippel 1988). Eric von Hippel has shown that firms that develop effective organisation tools for allowing customers to input potential new opportunities and solutions commonly outperform firms that do not. Such procedures for inputting customer ideas are often resisted as they increase uncertainty and make managerial tasks more complex.

Also critical in problem definition is that no system can do all things equally well, and the most effective way to organise will depend on which tasks the organisation seeks to perform. The first step in the development of an effective problem-solving system is to ask: what value proposition does the system want to offer to its customers; how does it seek to compete? A wide range of bases of competition might be feasible, including price, quality, flexibility, speed of response and innovativeness. Most firms tend to prioritise and emphasise one form of competition above others. The choice of competitive differentiation, or industry position, will then suggest choices along various dimensions of innovation system design and management.

Trade-offs are thus essential in innovation-oriented organisational systems, as in technical systems. In addition, managerial attention is limited. No manager can devote equal attention to more than a few tasks. Every effective innovation system selects a limited set of tasks to perform and on which to attempt to excel. Innovation managers must focus their systems in ways that are consistent with their organisation's competitive goals and with other internal elements of the system. Finally, many firms often allow focus to erode, even after strategic decisions have been made about which tasks to perform. Successful firms attract interest and are invited to accept new tasks and projects, which may be beyond their organisational capabilities and configuration. Complexity continually reappears.

Selecting Solution Options

Once a problem or opportunity has been selected, decisions must be made about which potential solutions are to be tested. To choose the set to be searched, organisations usually rely on previously acquired knowledge (Nelson 1982). The set of potential solutions must be limited. Indeed, the testing process will be quicker and more efficient the more focused the set.

Organisations accumulate knowledge to identify feasible solution sets by building experience in relevant fields (Levinthal and March 1981). As firms gain experience and accumulate relevant knowledge, their search effectiveness and option selection processes improve. Experience can yield at least three categories of useful knowledge: (a) insight into which problems are most valuable to solve, (b) deeper understanding of search tools and (c) information about where solutions are most likely to be found. Here again, the choice of organisational form, whether centralised or decentralised, will influence how this set is constructed. A centralised organisation is more likely to choose options that can readily be transformed into explicit data; a decentralised organisation will be more likely to seek options that rely less on cross-module integration.

Testing and Eliminating Unsuitable Options

There are many approaches to option testing. Tools for managing various types of uncertainty include: forecasting (attempting to predict the future); experimentation (testing several feasible futures); process control (removing sources of uncertainty); commitment (bet on a single future, perhaps thereby helping actually to bring it into being); variation and selection (allow variance and choose "good" outcomes); and flexibility (develop the capability to adapt to whatever the future eventuates; see West 2000).

Two tools for testing and eliminating options are experience and experimentation. Experience acts to narrow the set of solution options; experimentation acts to select among the elements of the set. Firms face a choice as they attempt to improve their problem-solving performance. They can build experience or they can increase experimentation capacity. These choices are not mutually exclusive, but in practice firms favour one or the other path (West and Iansiti 2003). While some firms rely upon their experience in the underlying science and technology or the context where the solution will be applied (for example, in manufacturing) for testing and eliminating options, others rely on experimentation for search capability and to test between options. The experience strategy, if successful, reduces cost and time in experimentation, making the problem-solving process more efficient; the experimentation strategy enables a wider search and faster option elimination. The experimentation strategy is strongly correlated with some measures of technical performance (West and Iansiti 2003). While projects with high levels of complexity gain from organisational processes that integrate experience in the project domain (for example, in manufacturing or banking services), projects with high levels of uncertainty benefit from wider experimentation (Allen, Tushman and Lee 1980). As organisations find over time one approach yields better results, they will tend to favour that approach and the associated tools used for testing between options.

Transferring Solutions for Implementation

The difficult task of implementing the solution is the final step in the problem-solving process. The solution must be passed from the site of discovery to the site of application. Such transfer is often problematic. Key issues in technology transfer include the difficulties of transmitting tacit knowledge, differences in the character and capabilities of the transmitter and the receiver and differences in risk-bearing capability. Successful solution transfer frequently requires adaptation on both sides of the process (Leonard-Barton 1988).

Considerable research has explored the factors that account for superior performance in knowledge transfer and implementation. All can be viewed as means to reduce or cope with complexity and uncertainty. Better transfer is found when: there is a single, high-level ("heavy-weight") project manager (integration through centralisation; Clark and Fujimoto 1991); a dedicated project team is responsible for the project (Wheelwright and Clark 1992b); and a core of team members continues across different phases of the project (Keller 1986).

PROBLEM-SOLVING AS A STRATEGIC CHALLENGE

The problem-solving process thus presents both technical and administrative challenges, and those challenges constrain creativity. But there is a force that impels organisations toward the quest for innovation and its ingredient creativity, and that is competition. Competition with other firms introduces a strategic dimension to the organisation's choices, which calls for another level of creativity, tactical flexibility. All firms and most industries ultimately face challenges to their ability to attract and retain customers in the face of efforts by competitors to attract them away. And higher profits attract the attention of more potential competitors.

Innovation is a means to attract customers or block the efforts of others to win customers. Competition is often about price, and the drive for improved efficiency is an integral dimension of managing for competitiveness. But competition is based also on the technical characteristics of products—that is, on the services for which customers pay. On this dimension, competition is primarily technological. Firms compete by improving the performance characteristics and attributes of products, to differentiate their product lines from those of competitors. Innovative firms seek to create new product concepts and build "islands" of technological capabilities and specialisation that insulate them from imitation, and hence competition.

When firms succeed in this, they can appropriate the benefits of innovation. But firms cannot do this just once; they must create a flow of innovation over time for sustained profitability. Predicting—or creating—future market conditions and defining and building innovative capabilities to meet those conditions are creative problems.

The organisation's strategic response to competition can appear inefficient. Organisations sometimes undertake a wide range of apparently "inefficient" projects to support their strategic aims and thwart competitors. They undertake tasks that competitors find difficult to copy even if that means sacrificing efficiency. They offer greater product variety or maintain spare capacity to ensure quick response to customer demand. This may be inefficient along standard criteria. But such choices allow firms to achieve strategic differentiation and make their activities less easily imitable. Thus, the "best" strategy depends on the specific type of advantage the organisation is trying to achieve (Porter 1996). Another consequence of competitive learning is that the most effective arrangement of an innovation system changes over time. Learning implies that what worked last time may not be effective next time. Doing the unexpected or attempting more difficult tasks may thus be strategically valuable, even if not administratively compelling.

Effective planning for innovation must therefore proceed from the assumption that almost any competitive advantage in business can be overcome given sufficient time and effort, and any performance gap can be closed by a competitor, whether through organisational change, technology or increased effort. This is especially true for industries in which the basis of differentiation is knowledge.

In summary, organisations face increasing difficulties to maintain and integrate creativity due to the constraints imposed by their responses to complexity and uncertainty. But competition and strategic learning induce a strong force in the opposite direction. The survivors are those that succeed in innovating. How organisations manage this challenge can be seen as the essence of the challenge of innovation and creativity.

NOTES

1. Probably the best of the early work on these barriers remains Parkinson (1962).
2. In many industries, we have arrived at the future envisaged by Jaikumar: "The worker is likely to become completely separated from the physical elements of work—metal, lubricants and oil, executing procedures, and turning out parts. Work will, instead, become an act of conception, of creating new products and processes" (1988, 25).

3. Archaeologists have developed a similar analysis to explain the periodic implosion of complex societies throughout history and prehistory (Tainter 1990).

BIBLIOGRAPHY

Allen, T. J., M. L. Tushman and D. M. S. Lee. 1980. R&D performance as a function of internal communication, project management, and the nature of work. IEEE Transactions on Engineering Management 27: 694–708.

Arrow, K. J. 1974. The limits of organization. New York: Norton.

Austin, R., L. Devin and J. West. 2000. "Production as serial reconception". Presented at the Production and Operations Management Society Annual Conference, San Antonio, TX.

Baldwin, C. Y., and K. B. Clark. 2000. Design rules: The power of modularity. Cambridge, MA: MIT Press.

Bolton, P., and M. Dewatripont. 1994. The firm as a communication network. Quarterly Journal of Economics 109: 27–45.

Calvo, G., and S. Wellisz. 1978. Supervision, loss of control, and the optimal size of the firm. Journal of Political Economy 87: 943–52.

Clark, K. B., and T. Fujimoto. 1991. Product development performance. Boston: Harvard Business School Press.

Dosi, G., and L. Marengo. 1993. "Some elements of an evolutionary theory of organizational competences". In Evolutionary concepts in contemporary economics, ed. R. W. E. England, 211–35. Ann Arbor: University of Michigan Press.

Jaikumar, R. 1988. "From filing and fitting to flexible manufacturing: A study in the evolution of process control". Harvard Business School Working Paper.

Keller, R. T. 1986. Predictors of the performance of project groups in R&D organizations. Academy of Management Journal 29: 715–26.

Keren, M., and D. Levhari. 1979. The optimal span of control in a pure hierarchy. Management Science 25: 1162–72.

Leonard-Barton, D. 1988. Implementation as mutual adaptation of technology and organization. Research Policy 17: 251–67.

Levinthal, D., and J. G. March. 1981. A model of adaptive organizational search. Journal of Economic Behavior and Organization 2 (4): 307–33.

Milliken, F. J. 1987. Three types of perceived uncertainty about the environment: State, effect and response uncertainty. Academy of Management Review 12 (1): 133–43.

Nelson, R. R. 1961. Uncertainty, learning, and the economics of parallel research and development efforts. Review of Economics and Statistics XLIII: 351–64.

———. 1982. The role of knowledge in R&D efficiency. Quarterly Journal of Economics 97 (3): 453–70.

Newell, A., and H. A. Simon. 1972. Human problem solving. Englewood Cliffs, NJ: Prentice Hall.

Parkinson, C. N. 1962. In-laws and outlaws and Parkinson's third law. Cambridge, MA: Lorem Ipsum Books.

Porter, M. E. 1996. What is strategy? Harvard Business Review 74: 6–27.

Simon, H. A. 1962. The architecture of complexity. Proceedings of the American Philosophical Society 106: 467–82.

Skinner, W. 1974. The focused factory. Harvard Business Review (May–June): 113–21.

Tainter, J. 1990. The collapse of complex societies. Cambridge: Cambridge University Press.

Van Creveld, M. 1985. Command in war. Cambridge, MA: Harvard University Press.

von Hippel, E. 1988. The sources of innovation. New York: Oxford University Press.

West, J. 2000. Operations strategy: Overview. Boston: Harvard Business School Press.

West, J., and M. Iansiti. 1999. Technology integration: Turning great research into great products. Harvard Business Review 75 (3): 69–79.

———. 2003. Experience, experimentation, and the accumulation of organizational capabilities: An empirical study of the evolution of R&D in the semiconductor industry. Research Policy 32 (5): 809–25.

Wheelwright, S. C., and K. B. Clark. 1992a. Creating project plans to focus product development. Harvard Business Review 70 (2): 70–82.

———. 1992b. Revolutionizing product development. New York: Free Press.

Winter, S. G. 1987. "Knowledge and competence as strategic assets". In The competitive challenge: Strategies for industrial innovation and renewal, ed. D. J. Teece, 159–84. Cambridge: Ballinger.

13 Creativity and Innovation
An Educational Perspective

Erica McWilliam

This chapter maps the contribution of education to creativity and innovation. To do so, it explores a number of different domains of educational theorising and practice, and what they bring to twenty-first-century understandings of creativity and innovation as related fields. The chapter proceeds by way of providing definitional clarity and an historical overview, before moving to consider substantive contributions under the following headings:

- Creativity and Education of Different Kinds
- Creativity and Purpose: Humanistic versus Utilitarian
- Creativity and Education at Different Levels
- Creativity and Individual versus Social Process

Finally it considers some policy implications of all this for contemporary and future education.

A few words about the nature of education to begin. The matter of whether or not education ought to be regarded as a discipline is a debate that is ongoing, but it is not one that needs to concern us here. The important matter is the extent to which education—as either a field of sub-disciplines or as a coherent discipline—has contributed to, and been impacted by, the increasing attention being paid to creativity and innovation.

Because educators have always played such a significant role in social identity formation, any idea that captures the social imagination, and the attention of governments and business, cannot fail to mobilise educational advocates, practitioners and critics. Thus educators have not been immune to the call to "more creativity and innovation" that is now ubiquitous in discourses of enterprise and utility, nor have they been absent from the more recent work of rethinking creativity and innovation, including the new purposes to which this rethinking can be put. Indeed, educators have been among the most vociferous advocates—and critics—of the host of versions of creativity and innovation that now abound in this increasingly crowded discursive field.

CREATIVITY AND INNOVATION DEFINED

If, as John Hartley avers, creativity is "the process through which new ideas are produced" while innovation is "the process through which they are implemented" (2004, xi), then educators have been much more focused on the former than the latter, as will become evident in what follows. This is not to deny a strong vocationalist tradition in education, or the importance attached to value-adding learning outcomes, but to indicate that there have been long-term tensions for educators between a *dominant liberal tradition* of education as an all-embracing "process of living" (Dewey 1897/1983) and an *instrumentalist tradition* of seeing education more squarely in terms of a preparation for work and/or economic success. Advocacy of creativity is generally associated with the former tradition, while innovation, where it occurs at all, is more likely to be picked up in the latter as "education and training for innovation" (see Macdonald, Assimakopoulos and Anderson 2007), of relevance to business and the post-compulsory sector, but of little interest to progressive educators of children, unless in the form of technologies that can be adopted to improve teaching and learning.

HISTORICAL OVERVIEW

Traditionally, educators have aimed to inculcate the basic literacies (including numeracy), provide an introduction to the broad range of disciplines that make up the knowledge base and develop the skills and capacities needed to be a fully functioning member of the social world. Thus the mantra "helping them reach their full potential" is well entrenched in the lexicon of every contemporary educator.

We have seen in the last 150 years or so the broadening of these educational purposes from a narrow instrumental focus on education for the needs of the industrial state to a much more comprehensive set of aims. This has produced, and been produced by, a wide range of disciplinary discourses feeding into education—various strands of psychology, sociology, philosophy and history—spawning in more recent times different kinds of education: vocational education and training, curriculum theory, equity, disability and special learning needs, learning science and so on. Alongside these developments, we have seen a greater recognition of the importance of ages and stages in educational development, and thus the burgeoning of educational expertise in specialist levels of education like early years, middle years, post-compulsory and higher—and also adult—education.

All this makes for a very rich field of conceptualisation and contestation when it comes to considering when and how education has made its contributions to creativity and innovation.

CREATIVITY AND EDUCATION OF DIFFERENT KINDS

Creativity and Arts Education

Creativity (disconnected from innovation except by way of new teaching methods or workplace training) has been closely identified with the sub-disciplines of *arts education*, seen as particularly important in the early years of educational growth and development. Elliott Eisner is one of many educational writers who extol the virtues of arts education for making available multiple perspectives to think through complex issues and to make sound and ethical judgments. In *The Arts and the Creation of Mind*, Eisner (2002) insists on the centrality of the arts in exploring the fullness of the world and providing an education that meets the spiritual, emotional, psychological and physical needs of all students.

Understanding creativity as fundamentally about arts education has been both a strength and a limitation when it comes to taking creativity seriously after primary schooling. Indeed, the further a student moves away from the early years, the more likely it is that creativity is perceived as garnish to the core English, maths and science roast, and also more distant from the "real world" of paid employment. As a curriculum component, "creative" is invariably followed by the word "arts", with arts being framed in turn as the lesser half of science–arts or work–play binaries.

When creativity does reappear as a curriculum possibility after the early years of schooling, it is most likely to be in the context of *remediating school failure* and/or social disadvantage, artistic expression being hailed by social workers and remedial teachers as a means by which "at-risk" young people might be productively re-engaged with economically productive work. This recent move to re-engage disaffected youth by means of art-based learning opportunities incorporates notions of literacy that go beyond traditional genres, with new research seeking to explore ways of using digital tools to allow young people to co-create forms that are unique to their experiences. The edited article, *Media, Learning, and Sites of Possibility* (Hill and Vasudevan 2008) and Jeffery Duncan-Andrade and Earnest Morrell's *Popular Culture and Critical Media Pedagogy in Secondary Literacy Classrooms* (2005) are both reflective of the interest of "critical" educators in using creative arts–based practices to link cultural, social and cognitive dimensions of learning better to serve disaffected adolescents. In this way creativity has come to take on the role of "rescuer" of those whom schooling has failed, a two-edged sword that accords creativity special powers in the remediation of school failure, while locating it at an even greater distance from academic norms.

Creativity and Thinking Skills

Alongside creativity as an *arts education platform* for early years development and a means for *remediating adolescent failure*, there has been a steady growth of interest in the extent to which *creative thinking* can be harnessed as a set of learnable cognitive skills that will produce better learning outcomes in both formal and informal educational contexts. Tangible evidence of such interest can be seen in the growing popularity of the Reggio Emilia schools that were set up in post-war Italy and continue to thrive in many Western countries. The "Reggio" vision of the child as naturally developing into a competent learner places the child in the centre of educational purposes and seeks to ensure that learning makes sense from the child's point of view. Thus opportunities for creative—i.e. *child-initiated*—engagement in the arts are central to the Reggio Emilia curriculum. Italy was also the birthplace of alternative thinking about a methodology for enhancing creativity in children by way of the Montessori method, based as it is on theories of child development originated by Italian educator Maria Montessori in the late nineteenth and early twentieth centuries. Like the approach of the Reggio Emilia schools, the Montessori method places emphasis *on self-directed activity* on the part of the child and clinical observation on the part of the teacher. Child-generated engagement with the physical environment is understood to be the means by which academic concepts and practical skills are fostered.

Edward de Bono has been highly influential in the emergent post-war field of creative thinking and the direct teaching of thinking as a skill (see De Bono 1995), although his impact on educational systems has been much more modest, given his ambivalence about assessment or evaluation. Others, however, have been more sanguine about continuing cognitive research into creative thinking by exploring the mental models and cognitive practices that seem to be most useful to the creativity project. Some of this literature reflects the tendency to conflate "creative" thinking with "critical" thinking, presenting them as either synonymous or so closely related as to be almost indistinguishable. Susan Wilks's *Critical and Creative Thinking: Strategies for Classroom Inquiry* (1995) is one example of this tendency, as is Sandra Menssen's (1993) "Critical Thinking and Construction of Knowledge", and Karoline Lrynock and Louise Robb's (1999) "Problem Solved: How to Coach Cognition".

More recently, there has been a reaction against the tendency to conflate "creative" thinking and "critical" thinking. At the same time cognition-based practitioners have insisted on the value of both "critical" and "creative" skills in our thinking. Critical thinking has been described within this tradition as "a matter of thinking clearly and rationally" while creative thinking consists in "coming up with new and relevant ideas". Creativity has been then further divided into two types—cognitive creativity and aesthetic creativity, the former of which is considered by many to be ripe for systematic teaching (see, for example, http://philosophy.hku.hk/think/).

Creativity as an Outcome of Teaching

The burgeoning interest in creativity as an *outcome of specific teaching methods* is both an outcome of new imperatives in professional work, and a response to evidence about the different ways that young people now learn (Hartman, Moskal and Dziuban 2005; Seely Brown 2006). A recent report issued by the European University Association (EUA) directs the university sector to consider "creativity" as central to their research and their teaching:

> The complex questions of the future will not be solved "by the book", but by creative, forward-looking individuals and groups who are not afraid to question established ideas and are able to cope with the insecurity and uncertainty that this entails. (EUA 2007, 6)

The problem, according to Norman Jackson, is not that creativity is absent from higher education but that it is omnipresent and yet not taken seriously as a generic approach; teaching-for-creativity is "rarely an explicit objective of the learning and assessment process" (2006, 4). In Europe, collaborative research projects have been recently set up to mitigate this problem (see Jeffrey 2006). Such research is pitted against a resilient tradition of transmissive pedagogy in higher education, with academics instructing passive students, who then regurgitate the "transmitted" discipline-based content for the purposes of assessment. Teaching-for-creativity continues to present a fundamental challenge to this teaching and learning culture.

However, some recent research is making inroads into this problem. McWilliam and Dawson's (2008) application of "boid" research is one example. They draw attention to the ways that research synthesising computer animation and biological behaviour can assist educators to optimise the creative capacity, building opportunities in the environments in which their students learn. The emphasis in this ongoing work (e.g. McWilliam, Dawson and Tan 2009) is on creativity as a *social* or team-based process, in particular as a learning outcome that can be fostered through better understanding of the mutually supportive processes through which productive space is made for "high-flying" learning challenges. It focuses not on an individual's brain power but their capacity to work strategically in a team of high-powered, self-managing learners. This is a learning disposition that is of increasing relevance to "symbolic analysts"—the imaginative and creative thinkers who build the capacity of an organisation to compete in a highly demanding economic environment through their capacity to:

1. Theorise and/or relate empirical data or other forms of evidence using formulae and equations but also innovative models and metaphors.
2. See the part in the context of the wider and more complex whole.
3. Intuitively or analytically experiment with ideas and their products.
4. Collaborate with others in ways that increase opportunities for successful innovation (adapted from Yorke 2006, 5).

With manufacturing and other routinised types of work disappearing in highly technologised workplaces, such skills are at a premium in the global marketplace, notwithstanding the economic turbulence of recent times and its negative impact on risk and experimentation. It is worth remembering that Google and Microsoft were both founded in the tough economic times of the 1970s.

CREATIVITY AND PURPOSE: HUMANIST VERSUS UTILITARIAN/INSTRUMENTAL

Nineteenth Century Legacy

Educational systems have been geared up, by and large, to produce routine thinkers for routine work. In the nineteenth century, mass schooling, the predominant means by which young people in Western countries were prepared for their roles in the social world, was designed to produce a citizen whose moral, intellectual and social habits were closely aligned with an industrial social order. Alvin Toffler, writing in his classic prophetic book, *Future Shock* (1970), half a century ago, sums this up succinctly:

> Mass education was the ingenious machine constructed by industrialism to produce the kind of adults it needed. The problem was inordinately complex. How to pre-adapt children for a new world—a world of repetitive indoor toil, smoke, noise, machines, crowded living conditions, collective discipline, a world in which time was to be regulated not by the cycle of the sun and moon, but by the factory whistle and the clock. (362)

With "a place for everything with everything in its place", industrial education exemplified the triumph of order and Protestant asceticism over serendipity, arbitrariness and self-indulgence. The factory model of schooling taught post-Victorian generations that a preparation for work involved learning to eschew temporary gratification and to tolerate, indeed welcome, repetitive and routine experiences in the expectation that these habits would lead to long-term job security, which in turn would bring economic and social prosperity. As the "raw materials" of the educational factory, children continue to be channelled into "streams"—academic, general, vocational—that may delimit their life chances in terms of future employment.

Humanism as a Challenge to Instrumentalism

Not all nineteenth-century educators, however, adopted this narrow utilitarian view of the nature and purposes of education. Many of the very important philosophical antecedents to creativity as an important contributor to the fully functioning social individual were laid down during the late nineteenth century, the most notable in the work of founding educational father, John Dewey. Dewey's focus was squarely on the human experience. He had a powerful sense of the importance of education in individual and social progress, and in the importance of understanding the power of art to engage and transform human experience. In "My Pedagogic Creed" (1897/1983, 239), Dewey pre-empted the humanistic and transdisciplinary moves that would become so important in the twentieth century and beyond. He stated:

> Education [conceived as social progress and reform] ... marks the most perfect and intimate union of science and art conceivable in human experience ... the art of thus giving shape to human powers and adapting them to social service is the supreme art; one calling into its service the best of artists; that no insight, sympathy, tact, executive power, is too great for such service.

In establishing this humanistic ideal as the central purpose of education, Dewey pre-empted much of the educational liberalism of the twentieth century, and with it the determination to "free" education from the industrial-model school. Experimental schools like A. S. Neill's Summerhill (established in England in 1921) derive much of their "romantic radicalism" from John Dewey, and this includes their determination to "free" the child from the "robotic" culture of mainstream schooling in order to allow their capacity as "creators" to flourish. In insisting on the centrality of creativity, Neill, drawing on Dewey, was as vociferous a critic of mainstream education for its lack of attention to creative play, as educational commentators like Ken Robinson (2001) are today.

Dewey's humanistic legacy is particularly evident in the later psychological studies of Abraham Maslow and Carl Rogers, both of whom provided a strong stimulus for advocating the importance of creativity to individual expression and inventiveness in a life well lived. These thinkers, as humanists, believed in the potential of each individual to be "self-actualised" as a fully and wholly human being. This belief represented a departure from both a behaviourist or Skinnerian tradition of understanding human beings as produced by outside or external forces, and a Freudian tradition of thinking focused on the *ego*, i.e. the interaction of primal drives (*id*) and the demands of community (*superego*).

Maslow's *Motivation and Personality* (1954) put in place the basic concepts for his later work *The Farther Reaches of Human Nature* (1971), in which he discussed the concept of creativeness as close to, even synonymous with, the concept of the healthy, self-actualising, fully human person. Maslow also made the connection between creative arts education and human fulfilment, not just for artistry. In understanding the creative person as immersed or absorbed in the present work of creating, his work has synergies with Mihaly Csikszentmihalyi's (1990) later concept of "flow", the mental state in which a person becomes fully immersed in the job at hand, becoming both disconnected from the external world and energised by the totality of the involvement with the task.

Clinical psychologist Carl Rogers (1951, 1969) was equally concerned to understand the process of becoming fully functioning or self-actualised.

Rogers's understanding of human experience, including his belief that all human beings have a natural potential for learning, has radical implications for conventional education, most of which he saw as subtly destructive in its effects on human experience. For Rogers, the "alternative" was a radical departure from teaching altogether:

> When I realize the implications . . . I shudder a bit at the distance I have come from the commonsense world that everyone knows is right . . . Such experience would imply that we would do away with teaching . . . examinations . . . grades and credits . . . degrees as a measure of competence . . . [and] the exposition of conclusions. I think I had better stop there. I do not want to become too fantastic. (1969, 257–58)

According to Rogers, creativity is one of five qualities exhibited when one is free from the constraints of conventional thinking and doing, the others being: openness to experience, existential living, organismic trust and experiential freedom. Creativity arises when a person feels both free and responsible, mobilising them to participate in a broader social world, either through contributions to the arts or the sciences. For Rogers, educational institutions militated against the experience of freedom and responsibility, and so his scholarship continued to frame creativity as invariably to be experienced "outside" formal education.

Rogers's notion of creativity is very close to "ages/stages" theorist Erik Erikson's (1964) concept of *generativity*, in that it is apparent when an individual is positively focused on the wider social world around them. For Erikson, *generativity* is the process of caring for the product of creative acts (i.e. one's own children), evident from about the age of twenty-five to age sixty-four, and acts as a mobiliser of the performance of caring tasks within the family unit and beyond it. (Its flip side is *stagnation*, or the inability to connect with or care for others.) Erikson proposes that those who are successful in crisis resolution, or overcoming barriers to transition from one "stage" of life to the next, develop positive ego strengths. However, while Erikson's psycho-social stages of development have been highly influential in education, those stages associated with adulthood are markedly less important to educational practitioners than the stages before adulthood. Thus *generativity* has not been taken up with anything like the same enthusiasm that is evident around Erikson's (1968) notion of *identity confusion*.

CREATIVITY AND EDUCATION AT DIFFERENT LEVELS

Early Childhood Education

Humanistic educational scholarship mobilised educators worldwide to find ways of working across the epistemological boundaries that had become so firmly fixed in formal education at all levels. In broad terms, the move from teacher-centred to child-centred and learning-centred educational principles has served creativity well, yet at the same time it served to split "child-focused" creativity from adult- and workplace-focused "innovation". One effect of this is that most "creativity" initiatives have been focused on *children*, and the means whereby their growth and development can be optimised. It is unsurprising, then, that creativity has been, in educational terms, very closely associated with the *imagination of the child*, and the activities that have the potential to enhance that imagination. Following from this, "innovation" has been evoked as a descriptor of improved teaching techniques and tools that make for "better" education—or more and better use of technology in education—rather than as an outcome of creativity.

In this framing, innovation *precedes* creativity rather than following it. This is a departure from most other conceptualisations of innovation as the product of creativity, something that is tangible evidence of novel idea adding economic value. Attempts to predict whether or not the creative talents of individuals *will* produce an innovation—e.g. Fiona Patterson's (2002) Innovation Potential Indicator—do not spring from, neither do they have an impact on, this child-centred notion of the importance of creativity. Indeed, there is strong resistance on the part of early childhood educators to any notion that their work as practitioners or researchers should be tied to economic ends. The language of enterprise—of "clients", "customers", "products" and so on—is anathema, almost universally, in this domain.

Arts education is the curriculum area most closely associated with child-centred creativity, and the domain where there is least debate, at least in terms of the importance of artistic endeavour to a child's growth and development. The *Handbook of Research on the Education of Young Children* (Spodek and Saracho 2006) makes it clear that the field has changed significantly since the publication of the first edition of this handbook in 1993, and that this has mobilised the editors to include a specific sub-section on the "Development of Creativity" within the theme of "Childhood Education and Child Development", thus acknowledging not just the increasing importance being given to creativity in early years education, but also the extent to which it is understood to be closely tied to larger issues of child development. Robert Schirrmacher's (2005) *Art and Creative Development for Young Children* is a good exemplar of creativity advocacy of this more recent type, incorporating as it does, in this revised version, an in-depth discussion of the ways that technology can assist teachers to foster children's visual art appreciation and production. It is also typical of its genre in that it provides practical activities and recipes to support in-service teachers.

Post-Compulsory Education

In more recent times, we have seen an expansion of the educational importance of creativity across the entire spectrum of the lifespan, from cradle to grave. The fact that university-graduate attributes have been linked to creative human capital has seen employers and policymakers alike looking to creativity, innovation and human talent as the engines of future productivity and social dynamism. The trend to value creative and relational capacities over narrow instrumental skills is also reflected in the UK, with employers seeking "multi-competent graduates" (Yorke 2006, 2) who have, "two sorts of high-level expertise: one emphasising discovery and the other focusing on exploiting the discoveries of others through market-related intelligence and the application of personal skills" (2006, 5).

Underneath these trends is a more fundamental assumption that productivity in the twenty-first century requires "a deep vein of creativity that is constantly renewing itself" (National Center on Education and the Economy [NCEE] 2007, 10). This sort of creativity is not limited to artists or youth or the creative industries, but includes all those employed now or in the future in a wide variety of endeavour, including computing, engineering, architecture, science, education, arts and multimedia. *All* university graduates, as potential future "creatives" (Cunningham 2006; Florida 2002), are argued to have a workforce future that is less focused on routine problem-solving and more focused on new social relationships, novel challenges and the synthesising of "big picture" scenarios. It is unsurprising, therefore, that of all the qualities employers are seeking in graduates, "imagination/creativity" are becoming much more prominent (The Pedagogy for Employability Group 2006).

CREATIVITY AND INDIVIDUAL VERSUS SOCIAL PROCESS

From Individual to Social

While creativity is expanding in terms of its relevance for different levels of education, it is also expanding in terms of its definition as both an individual and a social process. In broad terms, post-war psychological traditions tended to focus on individual cognitive capacity. However, there have been many different approaches to the study of creativity as a cognitive capacity. For example, *the relationship of creativity to intelligence* has been of much interest (see, for example, Guilford 1950; and also Gardner 1993, 1999), as well as *measurement of creative potential* (e.g. Guilford 1986; Torrance 1979) and the means by which to *augment creative behaviours* (e.g. Osborn 1953; Parnes 1967).

While debates about the nature of creativity continued through the 1980s and 1990s (see Sternberg 1988; Finke, Ward and Smith 1992), some consensus was appearing about the creative process as involving *the application of past experiences or ideas in novel ways.* Arthur VanGundy's Creative Problem Solving (CPS) Model (VanGundy 1987), based upon the work of both Alex Osborn and Sid Parnes, began to solidify creativity in terms of steps in a process, in this case, fact-finding, problem-finding, idea-finding, solution-finding and acceptance-finding. The cognitive skills they identified as underpinning such creative behaviour included fluency, flexibility, visualisation, imagination, expressiveness and openness (i.e. resistance to closure). This opened up the possibility for thinking of creativity beyond an individual cognitive capacity, i.e. as *a characteristic of personality, or an outcome of learning, or as situational.*

It was in the 1980s, then, that we saw the emergence of a now widespread acknowledgment of the significant role that *social processes* play in creative thinking and doing. Teresa Amabile's work (1983, 1996) was and still is highly influential in rethinking creative performance as enhanced by the social environment, task motivation, domain-related skills and creative-related processes, not simply by internal processes to do with the brain or individual nature. This shift provided a platform for later moves to pluralise and democratise the concept of creativity, and these in turn allowed it to be possible to "think" of creativity as *learnable*, and possibly, therefore, teachable. In other words, the descriptor "creative" is not reserved for the artistic few, but comes to connote abilities and dispositions that could be developed and nurtured through education.

These two traditions of thinking about the nature of creative processes—i.e. as either an individual process of intuitive, subjective ideation or a social process with generic applicability—are reflected in two "generations" of understandings held by teachers. These two dominant ways of understanding creativity are aligned with broader philosophical understandings about the nature of the human subject as, on the one hand, an individual whose talents arise "from within", and, on the other, a social being whose capacities are a product of their relations with others. In 2006–2007, McWilliam (2007) led an investigation into the prevailing views of creativity among award-winning academic teachers in Australia, comparing these views with those of academics surveyed in the 2006 UK's Imaginative Curriculum Project. While some minor differences were noted in the study responses, two attitudinal trends were evident. First, at least two-thirds of the academic teachers in both studies hold the view that creativity is a skill that can be fostered and nurtured through teaching. The second was that many teachers in both studies hold a mixture of "first-generation" (individualistic) and "second-generation" (social, pluralistic) understandings, with the latter being argued to be more useful in terms of developing precise teaching and learning strategies for building creative capacity (see McWilliam 2007, 3; Table 13.1).

While popular notions of creativity continue to reflect "first-generation" understandings (i.e. individualistic and internally driven), creativity is a capacity that is now being acknowledged by increasing numbers of educational scholars worldwide as an observable and valuable component of social and economic enterprise. Creative capacity—as the ability to produce ideas can be turned into valuable products and services—is coming to be regarded as fundamental to an increasingly complex, challenge-ridden and rapidly changing economic and social order. In Mihaly Csikszentmihalyi's

Table 13.1   First- and Second-Generation Creativity Concepts Held by Academic
Teachers

First-generation creativity concepts	Second-generation creativity concepts
"Soft", serendipitous, non-economic	"Hard" and an economic driver
Singularised	Pluralised/team-based
Spontaneous/arising from the inner self	Dispositional and environmental
Outside the box or any other metric	Requires rules and boundaries
Arts-based	Generalisable across the disciplines
Natural or innate	Learnable
Not amenable to teaching	Teachable
Not amenable to assessment	Assessable

Source: Adapted from McWilliam (2007, 4).

terms, creativity is *"no longer a luxury for the few, but . . . a necessity for all"* (2006, xviii). A further important perspective has been added through Csikszentmihalyi's insistence on *the community, not the individual,* as the unit that matters when seeking to foster creativity. This proposition challenges conceptions of creativity that are limited to individualistic psychological traits, and this has pre-empted a shift in scholarly interest *from the* creative individual to the creative, *dynamic team*, the latter being enabled by their internal social dynamics, i.e. their capacity to "flock together" (Dobrev 2005), and their robust social networks, to generate more creativity than they could achieve as separate individuals.

There is now a significant body of educational scholarship (e.g. Amabile 1996; Howkins 2001; Jackson et al. 2006; McWilliam and Dawson 2008; McWilliam and Haukka 2008) that has moved to unhook creativity from "artiness", individual talent and idiosyncrasy, and to render it an economically valuable form of human capital, team-based, observable and learnable. Some researchers, (e.g. Jackson), have been particularly interested in working at the nexus of creativity, pedagogy and policy, while others such as Amabile, and also Sternberg (1999), have focused on how creativity may be fostered through organisational climate and collaboration. These developments in educational thinking draw not simply on the disciplines of psychology and sociology, but also on organisational and business studies and scholarship emanating from the newly emergent "creative industries".

The Creative Industries

The "creative industries" are exemplary locations of new types of enterprise that exist for a wide range of purposes, from creative design and construction to innovation in the social and human services, and they include but are not limited to, media and the visual and performing arts. Like Richard Florida's (2002) "creative class", these industries are not concentrated in one sector or creative ghetto but are highly integrated across the economy, and they are highly innovative not only in terms of artistic expression, but in terms of business models, modes of organising, integration of technology and formulation of new products and services (Howkins 2001). Their stock-in-trade is the exploitation of symbolic knowledge and skills through artistic works and through adding value and marketing, combining commercial knowledge and application with aesthetic modes of knowing and doing.

The emerging importance of the creative industries in terms of global productivity has been a strong impetus to moving discussions of creativity away from first-generation to second-generation definitions. The latter build on understandings first made public over fifty years ago in Arthur Koestler's *The Act of Creation*, in which Koestler identified the decisive phase of creativity as the capacity to "perceive . . . a situation or event in two habitually incompatible associative contexts" (1964, 95)—the capacity to select, reshuffle, combine or synthesise already existing facts, ideas, images and skills in original ways. In a similar vein, David Perkins (1981) argued in *The Mind's Best Work* that skills like pattern recognition, creation of analogies and mental models, the ability to cross domains, exploration of alternatives, knowledge of schema for problem-solving, fluency of thought and so on, are all creative dispositions or cognitive habits that can and should be both taught and learned to enhance creative problem-solving, symbolic analysis and related thinking skills.

Creativity and/as Intelligence

Howard Gardner's *Intelligence Reframed: Multiple Intelligences for the 21st Century* (1999), augmenting as it does his earlier work, *Multiple Intelligences: The Theory in Practice* (1993), has been highly significant in terms of transdisciplinary reframing of intelligence in general, and this has been important to rethinking creativity in terms of its relationship to intelligence. Drawing on an earlier generation of thinkers about human capacity—namely, psychoanalyst Erik Erikson, sociologist David Riesman and cognitive psychologist Jerome Bruner—Gardner formulated a provisional list of seven intelligences: *linguistic intelligence* and *logical-mathematical intelligence* (most valued in school settings); *musical intelligence, bodily-kinesthetic intelligence and spatial intelligence* (most closely associated with the creative arts); and *interpersonal intelligence* and *intrapersonal intelligence*, both of which focus on the personal. Later research and reflection saw *naturalist intelligence, spiritual intelligence* and *existential intelligence* added to the list. He sees these intelligences as rarely operating independently of each other, but rather as working together in complementing ways that allow people to develop skills and solve problems. All Gardner's intelligences can be seen to be implicated, to a greater or lesser degree, in the building of creative capacity for twenty-first-century social and economic purposes. As social scientist Mark Warschauer (2007) points out, we live in paradoxical times in which both digital and traditional literacies are needed, and in which social inter-active competencies and moral-ethical sensibilities around sustainability are crucial to twenty-first-century citizenry. It is hard, therefore, to see which of Gardner's intelligences could be "lopped off" as unnecessary to creative civic participation.

Gardner's Theory of Multiple Intelligences is incorporated in David Perkins's Project Zero. Founded in 1967, Project Zero is a research and development group at the Harvard Graduate School of Education that investigates human intelligence, creativity, understanding and learning. It is a meta-project for broadening and deepening thinking capacity, reaching out to schools, cultural institutions, museums and the business world. For Perkins, a thoughtful learning culture is not achieved within the school walls but extends more broadly into culture and community. This insistence on broadening and deepening educational experience in order to improve thinking capacity has resonances with Marilyn Adams's *Odyssey: A Curriculum for Thinking* (1986), an experimentally validated five-volume classroom instruction series on thinking skills that was originally developed for students in the Republic of Venezuela.

Creativity, Wisdom, Trusteeship

A further, more recent example of how educational research into thinking, creativity and imagination blurs the boundaries of cognition, philosophy, sociology and ethics, is evident in the collection *Creativity, Wisdom and Trusteeship: Exploring the Role of Education*, edited by Anna Craft, Howard Gardner and Guy Claxton (2008). Their contention is that "a blend of creativity and wisdom combined with revisiting the notion of trusteeship . . . would be highly desirable, and perhaps even necessary, for the survival of the world as we know it and as we would like to see it" (2008, 1). Their insistence on a holistic and interrelated notion of creativity is very much a product of child-centred moves that were being made for a decade or more.

A theme that emerges strongly in the Craft, Gardner and Claxton collection is the insistence on the higher purposes of creativity being those of personal fulfilment and social betterment rather than utility or instrumentalism. This is a theme that, as indicated earlier, has been a perennial of educational discourse, continuing unabated to the present day. Writing in the aforementioned collection, Hans Knoop makes a plea for focusing on the "enabling virtues" of "creative wisdom and wise creativity" and is highly critical of the sort of the "unbalanced, even dangerous growth that shows how much of what is called *human creativity* would be better named as *human destructiveness*" (2008, 124; emphasis original). Knoop's critique is typical of the critical educational perspective that is pitted against any unconditional celebration of creativity's role in economic growth. He is one of a number of educators who express ambivalence about uncritical engagement with a "universalized creativity discourse" that is identified by its "social class-based assumptions", its "strong individuation and self-reliance", its "future orientation" and its insistence on "control over one's environment" (Craft 2008, 24).

In the same volume, Craft insists that the cultural saturation of creativity in the marketplace has the strong potential to "rub away cultural political and socioeconomic differences", leaving educators "doing nothing more than accepting and implementing the policy requirements and scaffolding offered to us as educational practitioners" (2008, 27). Put bluntly, where creativity is seen to be harnessed to economic productivity and only economic productivity, then it is more likely to be damned than praised by those who see themselves as needing to reassert the moral-ethical role of creativity as a sufficient and laudable educative end in itself. This is not a debate that will be won or lost—it will continue to be played out in the new literature that is emerging from and for a new century.

POLICY IMPLICATIONS

The theory and practice of education has come a long way in recognising the importance of creativity in the shaping of the life well lived. As we begin the second decade of the twenty-first century, we are seeing more recognition of the universality of creativity and more interest in investigating the ways creativity can be fostered at all educational levels, from early years to doctoral studies. However, there remains much to do if the rhetoric around creative capacity building is to become the reality of what schools and universities are actually committed to producing. The structures of formal education at all levels are proving to be highly change-resistant.

With all economies worldwide being buffeted by uncertainty in recent times, the lives of young people are fast becoming much more complex in terms of the issues they face and the choices that are available to them, and they will therefore need to exercise more creative thinking and activity in their daily lives. The call to "open up" education to the sort of risk-taking and experimental thinking that are markers of creativity cuts across the strong push for risk minimisation and standardised testing that are features of the current educational policy setting worldwide. The narrow performance-based logic of accountability that bedevils public education flies in the face not only of a broad agenda about distributive justice as a human rights issue, but also the logic of economic growth itself and the key role that creativity now plays in both of these spheres of human action and interest. Put another way, the risk-minimising culture of audit that is ubiquitous in our schools and universities is antithetical to the sort of risk-taking and experimentation that nourishes creative potential.

We anticipate that mainstream education will continue to be important because of the weight of evidence that a more highly educated population means more civic participation, more community infrastructure, a better lifestyle, a bigger pay-packet and a more productive economy. Schools and universities are likely to remain the key means by which young people move from basic literacy to the high levels of discipline-based literacy and numeracy needed to function optimally in a "super-complex" economic and social order. As expressed in a recent report from America's NCEE:

> This is a world in which a very high level of preparation in reading, writing, speaking, mathematics, science, literature, history, and the arts will be an indispensable foundation for everything that comes after for most members of the workforce. (2007, 6)

That said, there is also mounting evidence of the irrelevance of mainstream education to new modes of knowledge production and distribution, much of which is being made possible through the affordances of new digital technologies. According to economist John Quiggan (2007), these new modes of production are predominantly household driven, and thus connote a very different sort of "prod-user" identity from the consumer passivity that characterised households in the last century. Digitalisation makes enterprise much less about routinised labour, centrally located offices and 9-to-5-ism, and much more about understanding, developing and maintaining the creative dispositions and conditions that people need to turn symbolic knowledge into economic and social assets. The "creative industries" are exemplary locations of this type of enterprise. They are industries that exist for a wide range of purposes, from creative design and construction to innovation in the social and human services. Their stock-in-trade is the exploitation of symbolic knowledge and skills, not only through artistic works but also through value adding and marketing. In this sense, they combine commercial knowledge and application with aesthetic modes of knowing and doing in ways that mainstream education does not, given its heavy reliance on disciplinary boundaries, streaming and lock-step progression.

If education is to address the gap between "creativity" rhetoric and the dominant pedagogical reality, policymakers, educational leaders and practitioners will (all) need to focus (more squarely) on the following:

1. Ensuring that creativity is not (just) limited to arts education in the curriculum but is also directed to the construction of borderlands between and among disciplines for solving complex problems.
2. Ensuring that creativity (as both epistemological agility and artistic expression) is explicitly fostered from the early years to doctoral studies. In other words, working to ensure that creativity infuses the

disciplines of sciences, mathematics and language learning, not simply the arts.

3. Building collaborative teaching and learning cultures in ways that build creative capacity at all levels of education, and aligning curriculum pedagogy and assessment towards this end.

4. Building the creativity/innovation nexus through closer alliance with employing bodies. Such partnerships are not easily built or sustained, but public/private consortia are beginning to emerge to fill the gap between educators and enterprise. Environmental sustainability imperatives are forcing both ends together in some cases: robust critical engagement with new knowledge, new cultural forms and new modes of expression. For example, allowing space in the curriculum for working in "design mode", not just disciplinary or "truth mode". This means suspending "the facts" temporarily and explicitly in order to encourage young people to think, as many designers do, outside their cultural logics to find creative ways of engaging with familiar questions or dilemmas. This needs to be part of education's core business, not simply a fringe or marginal activity.

5. Conducting/undertaking empirical research into the nature and value of creative capacity. This means exploring appropriate ways of characterising the value-addedness of creativity for living, learning and earning purposes. Recent studies such as those documented in McWilliam, Dawson and Tan (2009) are examples of how this might be done, including the uses that can be made of digital tools for methodological purposes and pedagogical diagnosis.

In summary, much remains to be done when it comes to theorising or investigating how ideas are translated into innovative actions, processes and products. As the history of education's engagement with creativity makes clear, this will be most usefully done through multiple approaches that can draw on other relevant social scientific traditions, rather than seeking to maintain rigid disciplinary distinctions, or to tolerate mediocre schooling in which "low threat, low challenge" is the norm. Creativity, demanding as it does a high challenge, risk-taking disposition, promises the pleasure *and* the rigour of original thinking and doing. This is not about doing away with traditional educational expertise, nor is it about ignoring educators' duty of care or simply "going digital". It is about ensuring that education moves closer to the "low threat, high challenge" learning culture that is most relevant to building creative capacity in and for the next generation.

BIBLIOGRAPHY

Adams, M. 1986. *Odyssey: A curriculum for thinking. Reading Research and Education Center, University of Illinois. Learning, development, and conceptual change series*. Urbana-Champaign, IL: Charlesbridge Press.

Amabile, T. M. 1996. *Creativity in context.* Bolder, CO: Westview.
———. 1983. *The social psychology of creativity.* New York: Springer-Verlag.
Craft, A. 2008. "Tensions in creativity and education". In *Creativity, wisdom and trusteeship: Exploring the role of education,* ed. A. Craft, H. Gardner and G. Claxton, 16–34. Thousand Oaks, CA: Corwin Press.
Craft, A., H. Gardner and G. Claxton, eds. 2008. *Creativity, wisdom and trusteeship: Exploring the role of education.* Thousand Oaks, CA: Corwin Press.
Csikszentmihalyi, M. 1990. *Flow: The psychology of optimal experience.* New York: Harper and Row.
———. 2006. "Foreword: Developing creativity". In *Developing creativity in higher education: An imaginative curriculum,* ed. N. Jackson, M. Oliver, M. Shaw and J. Wisdom, xi–xvii. London: Routledge.
Cunningham S. 2006. *What price a creative economy?* Platform Papers: Quarterly Essay on the Performing Arts 9 (July).
De Bono, E. 1995. *Serious creativity.* New York: HarperCollins.
Dewey, J. 1897/1983. "My pedagogic creed". In *Rethinking education: Selected readings in the educational ideologies,* ed. W. F. O'Neill, 232–39. Dubuque, IA: Kendall Hunt.
Dobrev, S. D. 2005. Career mobility and job flocking. *Social Science Research* 34: 800–820.
Duncan-Andrade, J., and E. Morrell. 2005. Popular culture and critical media pedagogy in secondary literacy classrooms. *International Journal of Learning* 12 (9): 273–80.
Eisner, E. 2002. *The arts and the creation of mind.* New Haven, CT: Yale University Press.
Erikson, E. H. 1964. *Insight and responsibility.* New York: Norton.
———. 1968. *Identity: Youth and crisis.* New York: Norton.
European University Association. 2007. *Creativity in higher education: Report on the EUA creativity project 2006–2007.* Brussels: The European University Association.
Finke, R. A., T. B. Ward and S. M. Smith. 1992. *Creative cognition.* Cambridge, MA: Bradford/MIT Press.
Florida, R. 2002. *The rise of the creative class.* New York: Basic Books.
Gardner, H. 1993. *Multiple intelligences: The theory in practice.* New York: Basic Books.
———. 1999. *Intelligence reframed. Multiple intelligences for the 21st century.* New York: Basic Books.
Guilford, J. P. 1950. Creativity. *American Psychologist* 5: 444–54.
———. 1986. *Creative talents: Their nature, uses and development.* Buffalo, NY: Bearly Ltd.
Hartley, J. 2004. "Preface". In *Innovation in Australian arts, media and design,* ed. R. Wissler, B. Haseman, S. Wallace and M. Keane, xi–xxi. Flaxton, ND: Postpressed.
Hartman, J., P. Moskal and C. Dziuban. 2005. "Preparing the academy of today for the learner of tomorrow". In *Educating the net generation: An Educause e-Book,* ed. D. G. Oblinger and J. A. Oblinger, chap. 6. www.educause.edu/educatingthenetgen.
Hill, M. L., and L. Vasudevan, eds. 2008. *Media, learning, and sites of possibility.* New York: P. Lang.
Howkins, J. 2001. *The creative economy: How people make money from ideas.* London: Allen Lane.
Jackson, N. 2006. "Imagining a different world". In *Developing creativity in higher education: An imaginative curriculum,* ed. N. Jackson, M. Oliver, M. Shaw and J. Wisdom, 1–9. London: Routledge.

Jackson, N., M. Oliver, M. Shaw and J. Wisdom, eds. 2006. *Developing creativity in higher education: An imaginative curriculum.* London: Routledge.

Jeffrey, B, ed. 2006. *Creative learning practices: European experiences.* London: The Tufnell Press.

Knoop, H. H. 2008. "Wise creativity and creative wisdom". In *Creativity, wisdom and trusteeship: Exploring the role of education,* ed. A. Craft, H. Gardner and G. Claxton, 119–32. Thousand Oaks, CA: Corwin Press.

Koestler, A. 1964. *The act of creation.* New York: Macmillan.

Lrynock, K., and L. Robb. 1999. Problem solved: How to coach cognition. *Educational Leadership* 57: 29–32.

Macdonald, S., D. Assimakopoulos and P. Anderson. 2007. Education and training for innovation in SMEs: A tale of exploitation. *International Small Business Journal* 25: 77–95.

Maslow, A. H. 1954. *Motivation and personality.* New York: Harper.

———. 1971. *The farther reaches of human nature.* New York: Viking Press.

McWilliam, E. 2007. "Carrick associate fellowship report". http://www.altc.edu.au/resource-developing-pedagogical-models-qut-2007.

———. 2008. *The creative workforce: Preparing young people for high flying futures.* Sydney: UNSW Press.

McWilliam, E., and S. Dawson. 2008. Teaching for creativity: Towards sustainable and replicable pedagogical practice. *Higher Education* 56 (6): 633–43.

McWilliam, E., S. Dawson and J. Tan. 2009. From vaporousness to visibility: What might evidence of creative capacity building actually look like? *UNESCO Observatory, Refereed E-Journal on "Creativity, Policy and Practice Discourses: Projective Tensions in the New Millennium"* 1 (3). http://www.abp.unimelb.edu.au/unesco/ejournal/vol-one-issue-three.html.

McWilliam, E., and S. Haukka. 2008. Educating the creative workforce: New directions for schools and universities. *British Educational Research Journal* 34 (5): 651–66.

Menssen, S. 1993. Critical thinking and construction of knowledge. *American Behavioral Scientist* 37: 85–93.

National Center on Education and the Economy. 2007. *Tough choices or tough times: The report of the new commission on the skills of the American workforce. National Center on Education and the Economy.* www.skillscommission.org.

Osborn, A. F. 1953. *Applied imagination.* Rev. ed. New York: Scribners.

Parnes, S. J. 1967. *Creative behavior guidebook.* New York: Scribners.

Patterson, F. 2002. Great minds don't think alike? Person level predictors of innovation at work. *International Review of Industrial and Organisational Psychology* 17: 115–44.

Pedagogy for Employability Group, The. 2006. "Pedagogy for employability". In *Learning and employability series one,* ed. M. Yorke, 1–44. York: The Higher Education Academy.

Perkins, D. 1981. *The mind's best work.* Cambridge MA: Harvard University Press.

Quiggan, J. 2007. "Innovation begins at home". Paper presented at the Digital Literacy and Creative Innovation in a Knowledge Economy Symposium, 29–30 March, Queensland State Library, South Bank.

Robinson, K. 2001. *Out of our minds: Learning to be creative.* Oxford: Capstone.

Rogers, C. R. 1969. *Freedom to learn: A view of what education might become.* Columbus, OH: Charles E. Merrill.

———. 1957/1983. "Personal thoughts on teaching and learning". In *Rethinking education: Selected readings in the educational ideologies,* ed. W. F. O'Neill, 255–58. Dubuque, IA: Kendall Hunt.

———. 1951. *Client-centred therapy: Its practice, implications and theory.* Boston: Houghton Mifflin.

Schirrmacher, R. 2005. *Art and creative development for young children.* Albany, NY: Delmar Thomson Learning.

Seely Brown, J. 2006. New learning environments for the 21st century: Exploring the edge. *Change* 38 (5): 18–25.

Spodek, B., and O. N. Saracho, eds. 2006. *Handbook for research on the education of young children.* Mahwah, NJ: Lawrence Erlbaum.

Sternberg, R. J. 1988. *The nature of creativity.* New York: Cambridge University Press.

———, ed. 1999. *Handbook of creativity.* Cambridge: Cambridge University Press.

Toffler, A. 1970. *Future shock.* London: Pan Books.

Torrance, E. 1979. *The search for Satori and creativity.* Buffalo, NY: Bearly Ltd.

VanGundy, A. B. 1987. *Creative problem solving.* New York: Quorum.

Wilks, Susan. 1995. *Critical and creative thinking: Strategies for classroom inquiry.* Portsmouth, NH: Heinemann.

Warschauer, M. 2007. The paradoxical future of digital learning. *Learning Inquiry* 1: 1–49.

Yorke, M. 2006. "Employability in higher education: What it is—what it is not". In *Learning and employability series one*, ed. M. Yorke, 1–44. York: The Higher Education Academy.

14 The Psychology of Creativity and Its Educational Consequences

John Sweller and Leon Mann

INTRODUCTION

Psychologists have long been interested in creativity. There is a large body of literature dating from the early twentieth century to be found in handbooks, encyclopaedias, books and specialist journals. The most recent *Annual Review of Psychology* chapters on creativity by Runco (2004) and Hennessey and Amabile (2010) demonstrate the wide range of topics of interest to psychologists and the subfields in which work in this area is organised. This work includes cognitive processes, intelligence, personality and individual differences including psychopathology, developmental aspects including lifespan changes in creative capacity and skills, instruction and educational aspects (see McWilliam, this volume), social and group factors (see Mann, Chapter 10 this volume), emotion and motivational factors and neuropsychological and neurological substrates in brain activity. However, cognitive skills and processes such as thinking, problem-solving, intuition and so-called "Eureka" moments are at the heart of the creativity enterprise for psychologists. Cognitive psychology underpins the descriptions and accounts of what is involved in creative thinking and activity, and how creative people differ—if at all—in their problem-solving and thinking processes. The interest in cognitive processes in creativity also underpins questions such as whether creativity is domain specific or general, why some people are more creative than others, whether all people are creative, the nature of creative genius and what is involved in exceptional, sustained creativity.

Sawyer (2006) identified a number of common myths that can be detrimental to understanding creativity: creativity comes from the unconscious; children are more creative than adults; creativity represents the inner spirit of the individual; creativity is a form of therapeutic self-discovery; creativity is spontaneous inspiration; many creative works go undiscovered in their own time and are only discovered decades later; everyone is creative; creativity is the same thing as originality; fine art is more creative than craft. It will be recognised that some myths are "pop" psychology, e.g. creativity comes from the unconscious; creativity is spontaneous inspiration. There is, therefore, an important role for psychological research to challenge and debunk erroneous myths.

A recurrent issue in the psychological study of creativity is whether it is domain general or domain specific (e.g. Sternberg 2005; Simonton 2007). The question is whether there is a generic creative process that transcends the particular problem-solving tasks of any given domain. Are people creative in one domain likely to be creative in other, unrelated domains? Is the creative thought of an Albert Einstein comparable with that of a Leonardo da Vinci? Or is there no such thing as a "one-size-fits-all" creative process? Like most of psychology's debates, there are many sides to the debate over whether creativity is domain specific. One side argues that creativity is a domain-general trait: people creative in one area are likely to be creative in other areas. The domain-general approach is associated with the psychometric study of individual differences (e.g. Plucker 2005). An opposing side argues that creativity is domain specific: people have islands of creativity, not a diffuse tendency to be creative. Sternberg (2005) pointed out that children who are unusually good in one domain (music, painting, science, etc.) are usually not especially good in others. Simonton (2004) also made the case for domain-specific creativity but in terms of different cognitive processes used across domains. He argued, for example, that artistic and scientific creativity are different because logic is a requirement for all scientific creativity, but not for most artistic creativity. Sawyer (2006) and Weisberg (2006) also championed the domain-specific view.

In this chapter we support the domain-specific view—but based on the principle that deep domain knowledge is essential for someone to be creative in a sustained way in any discipline or field.

CREATIVITY AND INNOVATION

Psychologists have much more to say about creativity than about innovation and entrepreneurial activity. Thus, psychology has not examined the relationship between creativity and innovation. There are, however, opportunities for psychology to examine creative problem-solving in managing innovation and also examine why some "creative" types become successful "innovation" types. It is also inevitable that psychologists, because of their interest in studying individual differences, will begin to ask what makes some people more involved in and better at innovation and entrepreneurship than others.

About the Fundamental Role of Cognitive Architecture in Creativity

In this chapter we describe new conceptual formulations in the psychology of creativity based on human information-processing principles and evolutionary theory principles (Sweller 2003, 2009; Sweller and Sweller 2006). We will draw upon research findings about information-processing and knowledge production from cognitive psychology as well as from principles of Darwin's theory of evolution.

It is not unusual to apply Darwin's evolutionary theory for model-building in areas outside biological evolution. While primary Darwinism focuses on Darwin's theory of biological evolution, some theorists apply what is known as secondary Darwinism to explain evolution phenomena not directly related to biological evolution. Examples are the immunological theory of antibody production or the growth and development of the human brain or cultural evolution as is the case with Dawkins's (1989) "memes", units of selection in cultural evolution. The propositions in this chapter about cognitive architecture or structure, i.e. human information-processing systems and how they relate to learning, memory, thinking and creativity, have implications for understanding and fostering creative potential, and for instruction and education.

It should be acknowledged that an evolutionary perspective or approach to creativity—while not centre stage in the psychology of creativity—has been around for some time. Predecessors include Campbell (1960) and Simonton (1999), who modified and elaborated Campbell's hypothesis of blind-variation-and-selective-retention in creativity. There are others, such as Popper (1979) and Vandervert, Schimpf and Liu (2007), who also take an evolutionary perspective on creativity. Indeed, Darwin (1871/2003) himself was the first to apply his principles to areas external to biology, such as human thought. The interest in a model of creativity based on an evolutionary perspective is shaped to some extent by recognition that psychology has so far failed to produce a grand, overarching theory of creativity that pulls together a wide range of disparate findings in an increasingly unruly field. (See Hennessey and Amabile [2010] for a brief discussion of evolutionary approaches to creativity—and supporters and critics of the approach.)

Psychologists, and especially psychometricians, have been interested in and studied creativity for many decades (e.g. Guilford 1959; Torrance 1966). That interest has led to psychometric tests of creativity based largely on the extent to which humans are capable of generating novel ideas, with the generation of novel, useful ideas (divergent thinking) providing an operational definition of creativity. These tests of creativity have not been particularly helpful for a deeper understanding of the meaning of creativity or for designing theory-based, practical techniques able to enhance creativity. As an obvious example, educational and training programs designed to enhance creativity—as defined by improved performance on standard tests of divergent thinking and ideation—have in our judgment not reliably demonstrated growth in creative potential or performance. Indeed, we are critical of most research on creativity training because it suffers from flawed research designs: for example, no systematic variation in the amount of training, lack of a control condition to provide evaluation of treatment effects, failure to randomly allocate participants to program-treatment and control conditions, inadequate measures of creative performance and lack of follow-up to check for enduring benefits of training against modest transitory effects. While creativity training interventions and school programs are popular, their benefits for actual creativity in schools and workplaces are not supported by a body of evidence based on properly designed experiments. Randomised, controlled experiments are essential before *any* instructional intervention or program can be recommended.

We suggest that our limited success in finding procedures that enhance human creativity is due to a gap between our rapidly advancing knowledge of human cognitive architecture, defined as the manner in which cognitive structures are organised, and the bulk of research into techniques for enhancing creativity. In the past, most creativity research ignored human cognitive architecture. With recent advances in our knowledge of human cognition, it may be time to place creativity within a human cognitive architectural framework. By doing so, we may be able to increase our knowledge of the relevant processes that underpin creativity and provide a firm base for research into educational and training techniques that could enhance human creativity.

About Human Cognitive Architecture

Consider an information-processing system with the following three characteristics:

1. The system is able to create new information.
2. It can remember the information for subsequent use.
3. It can disseminate the information over space and time.

Many readers will recognise this information-processing system as the one we humans use. We create new information whenever we have a novel idea such as the invention of the electric light bulb. We remember how to make and use light bulbs. In conjunction with that memory, we have language and ways of communicating that enable us to disseminate the knowledge concerning light bulbs to other people in other places and even to other people in future generations. This information-processing system seems to encapsulate human cognition. While most of us do not invent light bulbs very often, we frequently do create more mundane artefacts or procedures. Devising a new recipe or a new route to get from point A to point B in a city frequently involves creating, remembering and disseminating information.

In fact, these three characteristics of a possible information-processing system describe a system much more important than human cognition. They describe the system that gave rise to human cognition: evolution by natural selection. For example, biological evolution created the human brain along with all other biological structures and functions; it is structured to remember how to produce human brains; it has procedures in place that permit it to continue producing human brains over physical space and time.

Described in this manner, both human cognition and evolution by natural selection seem to "process information" similarly. If so, our cognitive system may have, in some important ways, evolved to mimic the information-processing system used by biological evolution. In brief, the two systems may be analogous.

Human cognition and biological evolution belong to a class of systems we call *natural information-processing systems* (Sweller and Sweller 2006). These systems exist in nature and have a common underlying base. That base can be described by a set of principles that constitute the essentials of both human cognition and biological evolution. The mechanisms of human creativity along with the mechanisms that determine the much greater creativity of evolution by natural selection can be found within those principles and, accordingly, the core of this chapter is concerned with principles that govern the functions of natural information-processing systems.

Two Categories of Knowledge

Before considering the principles of natural information-processing systems, we need to distinguish between biologically primary and biologically secondary knowledge (Geary 2005, 2007). Biologically primary knowledge consists of information we have evolved to acquire. As a consequence, we can acquire large amounts of biologically primary information easily and automatically without conscious effort. Learning to listen to and speak a first language, recognise faces, interact socially with other people and, of particular importance for creativity, learning general problem-solving strategies that apply to a wide range of disparate problems are all examples of biologically primary information. Most people in a functioning society will acquire these kinds of information unconsciously, simply as a consequence of being immersed in their society. They will neither need to make a conscious effort to recognise faces nor will they need to be specifically taught because all normally functioning humans are biologically primed to acquire this information.

Biologically primary knowledge can be used to assist in the acquisition of biologically secondary knowledge. Advanced societies require their citizens to learn increasingly large amounts of information that, from an evolutionary perspective, is new to humans and which we have not evolved to acquire. Unlike biologically primary knowledge, information about, for example, solving quadratic equations is acquired consciously. It frequently requires considerable effort. Societies set up institutions such as schools, universities and colleges to assist in the acquisition of biologically secondary information. Everything that is taught in educational institutions is biologically secondary. Our societies require members to acquire biologically secondary information but because we have only needed that information relatively recently in our biological history, we have not evolved to acquire it automatically. Unless we put in place specific societal structures to teach, for example, reading and writing, most people in a society will not learn to read and write. In contrast, virtually everyone will learn to listen and speak their native language without the need for educational systems.

Biologically secondary information is general in that, unlike biologically primary information, many categories of information can be acquired. There are rules that determine the manner in which such general information is acquired. Those rules apply to any natural system that processes general information and those rules also provide the cognitive machinery of creativity.

Natural Information-Processing Systems

The logic that underlies natural information-processing systems is identical irrespective of the type of system. Thus, the processes of evolution by natural selection which are relatively well specified can be used to throw light on the processes of human cognition. There are several ways in which the logic of natural information-processing systems can be stated (Sweller 2003, 2004), but in this chapter we will use five closely interrelated principles (Sweller and Sweller 2006).

1. The information store principle. In order to function in complex environments, natural information-processing systems must be able to handle a wide variety of conditions. A very large store of information is required for this purpose. In the case of human cognition, long-term memory provides that store.

As our understanding of long-term memory has increased, our understanding of its importance to human cognition has also increased. Long-term memory is not a peripheral structure used to remember a largely unrelated collection of facts. Rather, it is central to human cognition. What we see, hear and think about is at least as much determined by what is stored in long-term memory as by what impinges on our senses.

Initial evidence for the centrality of long-term memory in higher cognitive processes came from the game of chess. De Groot (1965; updated de Groot and Gobet 1996) investigated the cognitive processes of chess grandmasters and weekend players. Why do chess grandmasters almost invariably defeat weekend players? What creative processes are used by chess grandmasters to almost always design a series of winning moves? The only difference between weekend players and grandmasters that de Groot could find was in memory of chessboard configurations taken from real games. If grandmasters are shown chessboard configurations for about five seconds and then asked to reproduce them from memory, they are at least 70 per cent accurate while weekend players rarely achieve 30 per cent accuracy. The difference only can be found for remembering configurations taken from real games. Chase and Simon (1973) had similar results but also found that if the chess pieces were placed randomly on the board, there was no difference between weekend and grandmaster players—both groups performed below 30 per cent accuracy.

Why are chess grandmasters able to remember board configurations from real games and why is this memory skill advantageous when playing chess? The answer goes to the core of human cognition. It takes ten years or more of practice to become a chess grandmaster. During this period, chess grandmasters are not learning new, clever, general problem-solving strategies. Rather, they are learning to recognise the huge number of chessboard configurations they are likely to encounter in future games. They store in memory the relevant configurations and, from experience, the best moves for playing each one. Simon and Gilmartin (1973) estimated that a chess grandmaster can recognise fifty thousand to one hundred thousand board configurations and knows the likely best moves for responding to each. Unlike weekend players, who must attempt to devise a good move from scratch, grandmasters know what the best move is.

While the skill of chess grandmasters is specific to chess, we can all develop similar skills in areas of memory and problem-solving we have practised for a long time. If you are presented with forty random English letters for five seconds and asked to reproduce them from memory, you are likely to have a very low success rate. If you are presented with the letters "the skill of chess grandmasters is specific to chess" and asked to remember them, you are likely to do so flawlessly. This skill is similar to that of a chess grandmaster, the only difference being the relevant domain. In areas of our own professional specialisation or interest, we accumulate an enormous knowledge base over many years, and this knowledge base constitutes the core of our special skill. As will be discussed in the following, that domain-specific core of knowledge or skill is essential to creativity.

2. The borrowing and reorganising principle. The bulk of biologically secondary knowledge held in long-term memory is acquired from the long-term memories of other individuals. We imitate other people, listen to what they say and read what they write. We engage in these activities in order to transfer information from other people's long-term memory to our long-term memory. For example, anyone reading this book is reading it in order to obtain information held in the long-term memories of the authors.

The information-acquisition process is constructive rather than reproductive. We do not obtain a "recording" of the information held in another person's long-term memory. Rather, we combine new information obtained from listening, reading or imitating with old information held in our long-term memory. The two sources of information combine to form a "schema"—a cognitive framework or concept that helps organise and interpret information. Thus, schema formulation/development involves the reorganising, as well as borrowing of new information. The function of cognitive schemas is to structure information held in long-term memory. Thus, for example, we have schemas that enable us to recognise the infinite number of shapes that constitute handwritten letters of the alphabet; we have higher-order schemas that enable us to read combinations of those letters to form words, phrases and sentences; lastly, we have very sophisticated schemas that allow us to associate schemas for the complex squiggles that constitute writing with other schemas that reflect the external world. These hierarchies of schemas are acquired for all areas of competence, such as the schemas for remembering and recognising configurations developed by chess grandmasters described in the previous section.

3. The randomness as genesis principle. Consider a person solving a *novel* problem. There are two basic "moves" or strategies available. Most commonly, because no problem is totally novel, information is available in long-term memory suggesting either an appropriate move or set of possible appropriate moves. If such knowledge is available, it is likely to be used. Sometimes, however, appropriate knowledge is not available because there are two or more possible moves with no information in long-term memory available to distinguish between them. When knowledge is not available to distinguish between multiple possible moves, the alternative "move" generator, "random generate and test", is used. "Moves" or options are randomly generated and then tested for effectiveness in advancing or progressing the current problem closer to a satisfactory solution. In differing combinations, knowledge and the "random generate and test" principle are the only "move" generators available when solving novel problems.

The "random generate and test" principle has the important feature that a judgment about the effectiveness of a move can only be made after the move has been made, either physically or mentally. A decision concerning the effectiveness of a move cannot be made in advance. Once made, the move can either be accepted or rejected depending on whether it succeeds in moving the problem toward a solution.

The "random generate and test" principle for problem-solving is the ultimate source of all human creativity. Once knowledge is acquired following successful problem-solving, it can be stored and used indefinitely by its creator or used by others. A new knowledge base is established that may permit new problems to be identified and solved. Evolution by natural selection relies on the same creative process as "generate and test" through successive mutations and provides the ultimate source of biological variation.

4. The narrow limits of change principle. When there is insufficient prior knowledge available to reduce the number of moves to be tested under the "random generate and test" principle to a manageable number, there is another piece of cognitive architecture available to reduce the number and complexity of novel moves to be considered, viz. "the narrow limits of change" principle.

Problem-solving occurs in a working memory processor that has two well-known characteristics: it is extremely limited both temporally and in capacity. We are unable to remember novel information for more than about twenty seconds without rehearsal (Peterson and Peterson 1959) and cannot hold in memory more than about four to nine items of novel information presented at the same time (Cowan 2001; Miller 1956). Depending on the nature of the information-processing, we usually process no more than about two or three items of information in working memory. These limits in working memory reduce the amount of unorganised information we can handle.

Why is working memory so limited? Assume a working memory that can process about three elements of information. By the logic of permutations, there are 3! = 6 possible permutations of three elements. Determining which permutations are functional in the absence of knowledge is straightforward when only six permutations must be considered. In contrast, consider a putative working memory that can process ten elements. There are 10! = 3,628,800 possible permutations of ten elements. It is unrealistic to try to solve a problem in which 3,628,800 possible solutions must be considered and monitored. There is likely to be no evolutionary advantage to a working memory that can process thousands of elements simultaneously (Sweller 2003). Thus, the "narrow limits of change" principle dictates that human creativity must occur in a working memory structure that is bounded by extreme limits.

5. The environmental organising and linking principle. The previous four principles are needed to provide an effective information store for thinking and problem-solving. But the ultimate purpose of such a store is to permit appropriate action in an external environment. The "environmental organising and linking" principle provides the required cognitive mechanism. As noted, working memory has severe limitations when dealing with novel information from the environment. Working memory also deals with information from long-term memory. The characteristics of working memory when dealing with familiar information from long-term memory are different from its characteristics when dealing with unfamiliar information from the environment. Information in long-term memory is managed by cognitive schemas or frameworks that help organise and interpret information rather than by the "randomness as genesis" principle. Schematically organised information in long-term memory does not need to have limits and so there are no apparent limits to the amount of information that can be transferred from long-term to working memory. As a consequence, the characteristics of working memory differ dramatically depending on whether it is dealing with novel information or well-known, organised information from long-term memory. Because of these characteristics of working memory, Ericsson and Kintsch (1995) coined the term "long-term working memory" to refer to working memory when dealing with information transferred from the long-term store.

The large amount of information that we can transfer from long-term to working memory allows us to function effectively in our environment.

The governing principle is the "environmental organising and linking" principle, which allows us to rapidly access information stored in long-term memory to enable us to take appropriate action in a familiar environment. In the "environmental organising and linking" principle we have the ultimate reason for the architecture or design of our cognitive system. The system builds and stores information in long-term memory that enables us to function in an enormous variety of situations and environments we encounter.

CREATIVITY

These five principles comprise a model of human cognitive architecture that can be used to analyse human creativity. We began with the assumption that human cognition, including creativity, is likely to have evolved in a manner similar to processes found in the rest of the natural world. Evolution by natural selection is a creative mechanism that has created the biological world. Furthermore, biological evolution is a well-known mechanism whose processes can be readily mapped onto human cognition. Those processes throw considerable light on human creativity.

The first point to note is logical rather than empirical. Let us assume, in accordance with the aforementioned cognitive architecture, that when making a decision there are only two basic procedures available to us: we must use some combination of both knowledge held in long-term memory and of a random generation and test procedure. In other words, the randomness as genesis principle and the environmental organising and linking principle provide the universe of possibilities. No other procedure has been identified and until and unless an alternative is found, we have no choice but to assume that there are no alternatives, especially since the randomness as genesis and environmental organising and linking principles are the only ones required by natural selection to create novelty. We know that all biological variation can be sourced back to random mutation. Evolutionary biologists emphasise the importance of random mutation because there is no other conceivable natural process by which biological novelty can be created. There is no natural higher authority to determine the effectiveness of a mutation prior to it occurring. It must first occur and then be tested for effectiveness. Similarly, without prior knowledge, we cannot determine the effectiveness of a novel problem-solving move prior to making the move, either physically or mentally. Once the move has been made, frequently mentally, we can determine its effectiveness.

Our analysis explains why attempts to teach creativity as a general skill or capability are ineffective because creativity by its nature requires novelty and because the principles of cognitive architecture described in this chapter govern the limits and capacity to be creative when dealing with novel problems and challenges in unfamiliar domains.

Of course, it is possible to increase domain-specific knowledge and that knowledge, in turn, is essential for creativity. By increasing knowledge in an area, we increase the likelihood of a "random generate and test" process providing us with a novel, useful, desirable procedure or object. Based on the cognitive architecture discussed in this chapter, creativity training programs will not enhance general creativity across a range of unrelated domains or areas from chess to painting to scientific discovery. The lack of a body of evidence supporting such forms of creativity may be telling.

"Random generate and test" may be a biologically primary skill that does not need to be taught any more than learning to recognise faces needs to be taught. If we have all acquired the basic mechanism of creativity easily and naturally very early in life, why do people vary so dramatically in creativity? The answer can be found in the information store principle. While the processes by which we create new information are constant between individuals, the knowledge store to which those processes are applied can vary immensely. Creative processes applied to a knowledge base that is common to large segments of a population are likely to yield results that are not seen as having substantial value. The same creative processes applied to a unique knowledge base may yield unique results. In contrast to the "random generate and test" process—the basic mechanism of creativity that we acquire as a biologically primary skill—biologically secondary knowledge is acquired relatively slowly and laboriously but it can be taught and learned and it is that biologically secondary knowledge that is the key to creativity. Creative people are knowledgeable people.

ALTERNATIVE CONCEPTUALISATIONS OF CREATIVITY

Research into creativity has a long history in psychology. That history has rarely included attempts to integrate work on creativity with theories of human cognitive architecture. As a consequence, research into creativity has frequently been viewed as peripheral to the major concerns of research psychologists with few reliable, mainstream empirical findings. The present chapter can be used to interpret previous work on creativity using an evolutionary perspective of human cognitive architecture.

As indicated in the preceding, the model advanced in this chapter is compatible with Campbell's (1960) "blind variation and selective retention" theory and Simonton's (1999) theory of creativity espoused in a book subtitled *Darwinian Perspectives on Creativity*. Campbell's and Simonton's theories recognised the necessity of the Darwinian concept of chance as the ultimate origin of all biological variation and hence ideational variation. In using Darwinian evolution as a template for human creativity, the ideas presented in this chapter overlap with Campbell's and Simonton's writing. Simonton (1999) emphasised the role of chance combinations of mental elements or ideas in creativity. The "randomness as genesis" principle advanced in this chapter is similar to Simonton. Simonton suggested most chance combinations of ideas do not result in stable, coherent new ideas and are lost. The few ideas that are stable can be retained, providing the origins of creativity. The creative process is complete once the ideas are developed into a form that can be communicated to and valued by others. Simonton used elements of Darwin's own life and of others to illustrate how a Darwinian model of creativity might function. Simonton discusses three models for explaining creative output: a simple cumulative advantage ("the rich get richer"), multiplicative influence (creativity is made up of several factors, such as intellect and motivation) and combinatorial explosion (the more concepts in a network, the more likely a useful recombination will occur). While Simonton's theory is not closely tied to human cognitive architecture, it is compatible with the principles of cognitive architecture presented in this chapter.

In a book titled *Creativity in Science*, Simonton (2004) followed the Darwinian perspective to argue that scientific creativity can be examined from four principal perspectives: logic, genius, chance (serendipity) and zeitgeist. His thesis is that all four can be integrated into a coherent theory of creativity in science but, following Darwinian principles, chance is the primary cause, the other factors serving to enhance or constrain the operation of a chance combinatorial process. In essence, the principal concept proposed by Simonton is that creative genius is mainly a product of random and possibly unrelated ideas combined in an interesting way through the combinatorial process.

Before Simonton, Gruber (1980) used a historical case study approach to analyse Darwin's progress in developing his theory of evolution by natural selection. Gruber applied Piaget's theory, which is concerned with stages in the development of conceptual thinking in children, to conceptual growth in adults. He emphasised the exceptionally long periods required by Darwin to develop his ideas and argued that what appear to be dramatic breakthroughs frequently are the culminations of a very long series of small changes. The "narrow limits of change" principle advanced in the present chapter is very similar to elements in Gruber's analysis. While Gruber's analysis of creativity is not closely connected to principles of human cognitive architecture it is compatible with the cognitive architecture model.

Langley et al. (1987) used computational modelling to develop a model of scientific creativity that is based on principles of human problem-solving. The computational model proved capable of deriving elementary laws of physics such as Kepler's third law of planetary motion (which states that the cube of a planet's distance from the sun is proportional to the square of its period). The computational model could derive this law simply by being presented with periods and distances of various planets.

Modern conceptions of human problem-solving place a heavy emphasis on human cognitive architecture and by using principles of human problem-solving in their model, Langley et al. (1987) fit the cognitive architecture approach. However, there are some known aspects of human cognitive architecture described earlier in this chapter that cannot be easily modelled using computational systems. Long-term memory is central to the cognitive system and so the direction of people's creativity is dependent on the contents of their long-term memory. While humans have huge long-term memory capacity, it is not yet possible to devise a computer program with a similarly large amount of information in long-term memory. The creativity generator of Langley et al.'s (1987) model is entirely dependent on appropriate information being placed into the model's equivalent of long-term memory. Relying on relevant information placed in a long-term memory store closely mimics human creativity. Unfortunately, a side effect of only allowing the model to demonstrate "creativity" in areas already known to humans is that the model cannot demonstrate creativity in novel areas, considered a prerequisite for creativity.

The cognitive architecture model described in this chapter is offered as a conceptual framework to bring together key features of creative behaviour and to dismiss some myths, e.g. creativity is pure inspiration or mostly perspiration and the proposition that creativity is generic across domains. Although uncommon, it is possible to be highly creative in two or even three domains, but that is because the individual has—akin to chess masters—painstakingly over time acquired deep knowledge and developed cognitive schemas and conceptual frameworks in the respective domains to form the platform for sustained originality and novel solutions to problems in the domain.

The model is not intended as a formula or recipe for creativity. While focusing on "hard-wired" aspects of cognitive architecture or "structure" it can coexist with other approaches that emphasie culture, personality, motivation, social and environmental factors, etc., in the development, maintenance and practice of creativity, as those factors also contribute although they are not the primary drivers. Hennessey and Amabile (2010, 577) wrote: "Mumford and Antes (2007) best summarized the state of the field when they called for caution to be applied in any attempt to account for creative achievement based on a single model of the kind of knowledge or cognitive processes involved".

EDUCATIONAL IMPLICATIONS

There are two persistent questions associated with creativity in educational settings: can training and instructional techniques be used to enhance creativity and, if so, how? These questions can be answered in light of the model of cognitive architecture that we have proposed.

All learning, including learning how to be creative, requires a change in long-term memory. What changes in long-term memory might facilitate creativity? Two possible answers are: knowledge of a particular domain such as mathematics, science, art or music (known as domain-specific knowledge) or knowledge of a set of cognitive strategies that will enhance creativity in general irrespective of the domain (domain-general strategies). Many educationists seem to favour domain-general strategies but, unfortunately, despite decades of work, it is difficult to find a strategy that enhances creativity in a wide variety of areas and that is supported by a body of empirical literature based on randomised, controlled experiments. The cognitive architecture described earlier explains why. If the randomness as genesis principle provides the basis of creativity, searching for strategies that enhance general creativity is likely to be difficult. At present, the data (or lack thereof) suggests strongly that enhancing creativity by attempting to teach domain-general strategies may have limited success. However, as indicated earlier, strategies that can enhance creativity in a specific knowledge or artistic domain can be productive.

While there are claims that attempts to teach and train cognitive skills involved in creativity such as divergent thinking and "brainstorming" are effective, there is in our judgment little basis for any claim that such programs reliably enhance creativity as a generic skill or produce sustained creative performance. The research studies purporting to show the real and enduring benefits of creativity training are flawed by weaknesses in experimental design, inadequate controls, non-random allocation of participants to treatment and control conditions, inadequate operationalisation and measurement of the variables defined as creative performance. (See Scott, Leritz and Mumford [2004] for a review of the literature on creativity training that takes a more generous view of its benefits.)

We can, however, enhance creativity by enhancing domain-specific knowledge. Under the domain-general viewpoint, the extensive mathematics knowledge base possessed by all creative mathematicians, for example, is frequently assumed to be essential for creativity in mathematics (one can hardly be creative in mathematics without knowing *any* mathematics) but not a *direct* source of creativity. The contrasting, domain-specific view is that an appropriate knowledge base is the *major* teachable/learnable factor governing development and refinement of creativity. That view flows directly from the cognitive architecture model described in this chapter. The development of teachable/learnable domain-general strategies able to enhance creativity await both the development of a suitable cognitive theory and a body of empirical evidence based on randomised, controlled experiments demonstrating effectiveness.

BIBLIOGRAPHY

Campbell, D. 1960. Blind variation and selective retention in creative thought as in other knowledge processes. *Psychological Review* 6: 380–400.
Chase, W. G., and H. A. Simon. 1973. Perception in chess. *Cognitive Psychology* 4: 55–81.

Cowan, N. 2001. The magical number 4 in short-term memory: A reconsideration of mental storage capacity. *Behavioral and Brain Sciences* 24: 87–114.

Darwin, C. 1871/2003. *The descent of man*. London: Gibson Square.

Dawkins, R. 1989. *The selfish gene*. 2nd ed. Oxford: Oxford University Press.

de Groot, A. 1965. *Thought and choice in chess*. The Hague: Mouton.

de Groot, A., and F. Gobet. 1996. *Perception and memory in chess: Heuristics of the professional eye*. Assen, the Netherlands: Van Gorcum.

Ericsson, K. A., and W. Kintsch. 1995. Long-term working memory. *Psychological Review* 102: 211–45.

Geary, D. 2005. *The origin of mind: Evolution of brain, cognition, and general intelligence*. Washington, DC: American Psychological Association.

———. 2007. "Educating the evolved mind: Conceptual foundations for an evolutionary educational psychology". In *Psychological perspectives on contemporary educational issues*, ed. J. S. Carlson and J. R. Levin, 1–99. Greenwich, CT: Information Age Publishing.

Gruber, H. 1980. *Darwin on man: A psychological study of scientific creativity*. 2nd ed. Chicago: University of Chicago Press.

Guilford, J. P. 1959. The three faces of intellect. *American Psychologist* 14: 469–79.

Hennessey, B., and T. Amabile. 2010. Creativity. *Annual Review of Psychology* 61: 569–98.

Langley, P., H. Simon, G. Bradshaw and J. Zytkow. 1987. *Scientific discovery: Computational explorations of the creative process*. Cambridge, MA: MIT Press.

Miller, G. A. 1956. The magical number seven, plus or minus two: Some limits on our capacity for processing information. *Psychological Review* 63: 81–97.

Mumford, M. D., and A. L. Antes. 2007. Debates about the "general" picture: cognition and creative achievement. *Creativity Research Journal* 19: 367–74.

Peterson, L., and M. J. Peterson. 1959. Short-term retention of individual verbal items. *Journal of Experimental Psychology* 58: 193–98.

Plucker, J. A. 2005. "The (relatively) generalist view of creativity". In *Creativity across domains: Faces of the muse*, ed. J. C. Kaufman and J. Baer, 307–12. Mahwah, NJ: Erlbaum.

Popper, K. 1979. *Objective knowledge: An evolutionary approach*. Oxford: Clarendon Press.

Runco, M. A. 2004. Creativity. *Annual Review of Psychology* 55: 657–87.

Sawyer, R. K. 2006. Explaining creativity: The science of human innovation. Oxford: Oxford University Press.

Scott, G. M., L. E. Leritz and M. D. Mumford. 2004. The effectiveness of creativity training: a quantitative review. *Creativity Research Journal* 16: 361–88.

Simon, H., and K. Gilmartin. 1973. A simulation of memory for chess positions. *Cognitive Psychology* 5: 29–46.

Simonton, D. 1999. *Origins of genius: Darwinian perspectives on creativity*. New York: Oxford University Press.

———. 2004. *Creativity in science: Chance, logic, genius, and zeitgeist*. Cambridge: Cambridge University Press.

———. 2007. "Creativity: Specialized expertise or general cognitive processes?" In *Integrating the mind: Domain general versus domain specific processes in higher cognition*, ed. M. J. Roberts, 351–67. Hove, UK: Psychology Press.

Sternberg, R. J. 2005. "The domain generality versus specificity debate: How should it be posed?" In *Faces of the muse: How people think, work, and act creatively in diverse domains*, ed. J. C. Kaufman and J. Baer, 299–306. Mahwah, NJ: Erlbaum.

Sweller, J. 2003. "Evolution of human cognitive architecture". In *The psychology of learning and motivation vol. 43*, ed. B. Ross, 215–66. San Diego: Academic Press.

————. 2004. Instructional design consequences of an analogy between evolution by natural selection and human cognitive architecture. *Instructional Science* 32: 9–31.

————. 2009. Cognitive bases of human creativity. *Educational Psychology Review* 21: 11–19.

Sweller, J., and S. Sweller. 2006. Natural information processing systems. *Evolutionary Psychology* 4: 434–58.

Torrance, E. P. 1966. *Torrance tests of creative thinking.* Princeton, NJ: Personnel Press.

Vandervert, L., P. Schimpf and H. Liu, 2007. How working memory and the cerebellum collaborate to produce creativity and innovation. *Creativity Research Journal* 19: 1–19.

Weisberg, R. W. 2006. *Creativity: Understanding innovation in problem solving, science, invention, and the arts.* Hoboken, NJ: Wiley.

15 Creativity Meets Innovation
Examining Relationships and Pathways

Leon Mann

Chapter 1 identified several factors bringing creativity and innovation closer conceptually and in practice. The factors include the quest by governments, regions, cities and industries to develop their innovation systems to enhance productivity and performance; the emergence of the knowledge economy and of new creative industries (such as design and media); the strategic direction of many leading companies toward high-tech innovation to fuel the industrial "engines of tomorrow" while recognising that low-tech industries can also be highly creative and innovative; and, finally, a growing recognition that the most creative minds are needed to suggest solutions for the major problems of environmental degradation, climate change, energy depletion, food shortage, poverty, disease and population explosion.

However, creativity and innovation are still treated as separate although related processes in the scholarly literature and in policy formulation. There are many reasons: some stem from a traditional view that creativity is mysterious, private, expressive, about the arts, difficult to manage and not really productive, while innovation is material, observable, instrumental, about science and technology, measurable, manageable and productive. The contrasting views are sustained by old myths such as creativity is a "rare gift" and newly constructed myths such as innovation is what businesses do when they decide to change direction (see discussion of Sawyer 2006; Berkun 2007; and others in Chapter 1). And, as noted in Chapter 1, the social science disciplines—psychology, education, social psychology, sociology, economics, economic geography, history, law, political science, management, policy studies—tend to focus on either creativity *or* innovation in accordance with their own core subject matter.

We are interested in greater integration of creativity and innovation for several reasons. The first is for scholars and writers to advance knowledge and ways of thinking about creativity and innovation. Creativity (new ideas, concepts, thinking) does not invariably lead to innovation, but it is the basis of every major innovation. Treating creativity and innovation together is a way of mapping the dynamic association between the two and identifying knowledge gaps and opportunities for new research. The second is for planners and managers to understand how best to nurture creativity and foster innovation across domains. In order to devise effective learning strategies, incentives and institutional support for creativity and innovation, it is helpful to recognise the critical points in the creativity–innovation continuum to offer intervention and assistance. The third is that understanding how creativity and innovation connect provides policymakers with a basis for directing public- and private-sector resources and attention toward specific domains and activities.

ASSUMPTIONS ABOUT INTEGRATING CREATIVITY AND INNOVATION

We propose the following steps to achieve greater understanding of the relationship between creativity and innovation.

Defining Terms

A first step is to agree on what constitutes creativity and innovation. If definitions are vague and rubbery it will be impossible to examine how the two processes are related. My definition of innovation might not match someone else's. Ideally, we need agreement so as to build a cumulative body of knowledge and advance understanding of how the two processes connect in different spheres and domains, such as in a particular industry or at a particular time in history. We are interested in broad definitions of creativity and innovation to encompass all domains of activity. The following are broad definitions: "creativity is coming up with ideas . . . innovation is bringing ideas to life" (Davila, Epstein and Shelton 2006, xvii); "Creativity is thinking up new things. Innovation is doing new things" (Levitt 1981, 96); creativity involves the "activity of generating novel ideas and thinking", while innovation is "the realisation and utilisation of new ideas and concepts in practice" (Mann, Chapter 10, this volume); creativity is "the ability to bring something into being" while innovation is 'the introduction of novelties" (Christie, this volume, but note creativity is "ability" under this definition).

Another group of scholars give definitions relevant for the domains of economics and business. Thus for Edqvist, "Innovations are new creations of economic significance. Innovations may be of various kinds, e.g., technological as well as organizational" (1997, 3). Mark Dodgson (this volume) defines creativity as "the origination of ideas and insights" and innovation as "their application to commercial advantage in new products, services and business models". Jonathan West (Chapter 12, this volume) defines creativity as an important aspect of problem solving—it is a source for the introduction and evaluation of novel approaches. He defines innovation as a "change to an economic process . . . the introduction to actual use of new products, services or production processes". Social innovation (e.g. new forms of governance, communal organisation, political systems, etc.) and cultural innovation (e.g. new forms of theatre, dance, painting, cinema, etc.) do not fit under these definitions.

Our horizon in this volume is creativity and innovation in business and beyond. It is entirely likely that the relationship between creativity and innovation will differ markedly across domains and settings. There will be many pathways. But that is a matter to be investigated, not assumed. Accordingly, I take a broad view and regard creativity as original, novel thinking and ideas, while innovation is application and implementation of original and novel ideas in new products, processes and practices (business) but also beyond business, as new social and political institutions, social arrangements and cultural forms and activities.

Identifying Level of Creativity and Kind of Innovation

A second step is to identify the level of creativity and kind of innovation to be connected in the relationship. Creativity can be conceptualised as a continuum from commonplace, everyday creative problem-solving (little-c) to rare and exceptional creativity (Big-C). In the mid-range is the creativity of very capable and experienced people who often together with others come up with new ideas and solutions.

We surmise that Big-C creativity is more likely to connect to major innovation. Middle-c creativity, characteristic of many organisations and R&D teams, will connect to incremental innovation—modest new and improved products. Little-c (everyday) creativity does not produce innovation, but connects through uptake and efficient use of the products of innovation, for example, in discovering quicker ways to use your cell-phone to write brief text messages, etc. The distinctions made between the three Cs of creativity oversimplify the picture, but levels of creativity and of innovation must be taken into account when mapping and testing connections between them. Minor creativity is most unlikely to produce major innovation. There are implications for schools in regard to working with little-c (everyday) creativity to provide a platform for developing middle-c and even Big-C creativity (see Morelock and Feldman 1999; also McWilliam, this volume).

Choosing a Set of Indicators

The third step is to suggest a set of indicators to identify and represent creativity and innovation. Indicators of creativity are a challenge because classifying new ideas and thinking as creative can be subjective and a matter for debate. Innovation, on the other hand, refers to items and outcomes that are real and observable. But there is a risk when classifying something as an innovation that the assumption follows that it must have been preceded by creativity. We must avoid the trap of circular reasoning when investigating the relationship between creativity and innovation. Here is a fanciful example from Thomas Edison's invention labs to help trace the connection between creativity and innovation. Edison held patents for approximately one thousand new inventions, including the electric light bulb. We can examine the connections between creativity and innovation in Edison's labs if we do the following. Assume that every one of Edison's inventions was an innovation (of course, some are more novel than others). Then examine Edison's lab books and records for evidence of the concepts, ideas and proposals that preceded work on developing and testing each concept. We have then a trace of how the creative idea is connected with the innovative product. Edison's lab books might of course show that many concepts did not go on to become inventions. That, too, is valuable for understanding connections.

The indicators of creativity and innovation vary from domain to domain. In the domain of business and economics, indicators of innovation include number of patents, new products, processes and services, and new and more efficient practices. For a city, indicators of innovation will include successful implementation of effective solutions for the city's water supply and transport system problems, new affordable housing developments and reduction in crime.

Examining and Mapping Connections between Creativity and Innovation

A fourth step is to catalogue a set of demonstrated and notional connections (pathways) between creativity and innovation. The term *pathway* signifies a trail between two points, but the path can be direct or indirect. The aim is to identify some prominent and potential pathways for exploration.

There are numerous plausible pathways connecting creativity and innovation. The different pathways are determined by many factors such as the *purpose* of the creative activity, e.g. whether authorised and resourced by a firm to develop a new commercial product, or commissioned by a patron (the Medici family?) to create something exceptional. The purpose might be to achieve a radical innovation, such as J. Robert Oppenheimer's research team, which produced the atomic bomb (the Manhattan Project), or an incremental innovation, such as Nokia producing a mobile phone with a small keypad. The particular *domain*, e.g. whether creativity is part of an artistic, philosophical, scientific or technological endeavour, will make a difference. Our search for connections between creativity and innovation will invariably take us to fields where the desired outcome (societal, system, firm) is new products in which there is commercial interest, but it could also be cultural or literary products, or new forms of social organisation and governance. The social context in which creative work is performed, e.g. as a solo activity, in a team or in collaboration with an industry partner, will also make a difference in the nature and strength of the connecting pathways.

PATHWAYS BETWEEN CREATIVITY AND INNOVATION

Hardly anyone has made a systematic study of the nature and variety of connections. Here, I list some plausible pathways illustrated with examples as a way of directing attention to the topic. There are implications for knowledge and policy formulation if some pathways more reliably link creativity and innovation. To plot pathways we need to locate the particular terrain or domain where creativity and innovation meet. The concept of innovation system (Freeman 1995; Edqvist 1997; West, chapter 2 this volume) is a useful beginning. To my knowledge there is no concept of "creativity system" to match the concept of "innovation system". Significantly, human potential and capacity appear to be the most important components of national innovation systems, suggesting that creativity is factored into the innovation equation. The components often recognised as most important in innovation systems are human capital (a well-educated and skilled workforce), strong support for public-sector research (which means fundamental and curiosity-driven research) and the free flow of ideas (cf. Furman, Porter and Stern 2002). The most significant differentiators of firm innovation and productivity reflect creative opportunity, for example, from "spillover" of knowledge through proximity to excellent research universities (Zucker, Darby and Brewer 1998). This, of course, means ready access by the firm to creative new ideas and concepts and also skilled graduates from local universities. All industries report some reliance on university research, but it is particularly important in high tech areas such as drugs, computers, semi-conductors and medical equipment (Fabrizio 2006). But while universities are a source of some radical innovations, innovative firms cast their net widely when looking for innovation.

In sum, analysis of innovation systems across nations, regions, cities, industry sectors, etc., tell a similar story—that the creativity-based components are crucial for innovation. This is a good beginning, but analysis of innovation systems is about identifying components, not tracing pathways between them. An innovation system map is not a GPS with directions for moving from C (creativity) to I (innovation) and perhaps back again. We explore different types of connection in the next section.

Creativity Precedes Innovation

The first pathway posits creativity (new ideas) as the antecedent to innovation (implementation). This is a linear model and presumes a direct path from creativity to innovation. The organisation of scientific R&D labs in the private sector (especially) and in the public sector is predicated on this model. Jonathan West (Chapter 2, this volume) points out: "In some instances, new science gives birth to new technology and commercial innovation. This is the simplest picture, and the one which advocates of more spending on science and education usually have in mind. Here, innovation is seen as the commercialisation of inventions made in scientific labs". In this volume, examples of the creativity-pushes-innovation pathway are studies of teams in organisational research labs (Mann) and teams in design organisations (Dodgson). Dodgson refers to a "Think-Play-Do" process in design organisations such as architectural firms, in which think (the creative idea) is "tested" and advanced by play (tinkering with the concept) *en route* to do (innovation and implementation). Mann's studies of product development teams in research organisations place creativity in the first phases of project work, leading, if all goes well, to innovation. There are somewhat different connections between creativity and type of innovation in research teams and product development teams. Creativity is essential in both kinds of team, but creativity in research teams is the source of major discoveries and the promise of radical innovation, while creativity in development teams is the foundation of product improvement and incremental innovation. Hence, the connection between creativity and innovation should be more apparent in fundamental research teams than in product development teams. A point of interest in the creativity–innovation process is that the "creatives" who begin the journey may lose interest in the next arduous steps of turning the creative idea or concept into a practical innovation, or they are not much good at it. Accordingly the "creatives" who begin the process often hand over their clever new idea to the production, manufacturing and marketing people to complete and implement as a new product. But as West (Chapter 12, this volume) indicates, innovation is more likely when at least a core of team members stay with the project throughout a sequence of overlapping phases from the first phase of idea generation to the final phase of production and distribution (Waldman 1994).

The creativity–innovation trajectory is affected by the purpose of the research. Donald Stokes (1997) refers to four quadrants in scientific activity according to whether its purpose is oriented towards basic understanding or practical application (or both). Creative work (original research) in each quadrant follows a rather different pathway to innovation. In Edison's quadrant (after inventor Thomas Edison), scientists and engineers conduct applied research so as to invent, patent and commercialise something new, e.g. an electric light bulb. The pathway between creativity and innovation in this quadrant is strategically guided and planned to be as direct and short as possible. In Bohr's quadrant (after physicist Niels Bohr) scientists perform fundamental, curiosity-driven research. They want to publish—they have no particular interest in commercialisation of their research. The pathway between creativity and innovation in this quadrant is neither planned nor expected. An example is the long, unexpected path from Einstein's theory of relativity to the invention of laser technology. In Pasteur's quadrant (after Louis Pasteur) scientists aim to advance scientific understanding and also provide real-world utility. This can be a tortuous path between creativity and innovation. In the fourth quadrant scientists muddle along with little aim or purpose. The pathway between any creativity and any innovation is haphazard and chaotic. In sum, even in the simplest model—a direct pathway from creativity to innovation—how the research is organised and for what purposes has an impact on how well the two are connected.

The creativity–innovation model applies to new theories and concepts across virtually every domain of knowledge. Consider Sigmund Freud's (1910) theory of the unconscious as an example of how new ideas about repressed impulses (creativity) gave rise to psychoanalysis, a new way of diagnosing and treating mental illness (innovation). The pathway model also applies to the historical analysis of new movements in political and social ideas as precursors to periods of great change and social innovation.

A variation of the model is that creativity in overlapping domains can trigger new pathways to substantial innovation. An example is the combination of ideas from computer science, medical science and statistics to produce the innovation of expert systems for diagnostic and treatment decisions in medical practice. Another example is the coming together of creative new ideas from psychoanalysis, anthropology, literature and art (surrealism) to produce innovation in cinema (see Jean Cocteau's films). A major rationale for multidisciplinary work in all fields but especially in science and technology is that creative work occurs at the intersections of several disciplines and can trigger breakthrough solutions for complex problems. When several disciplines are bubbling with creative new ideas, the innovation from their collaboration can be quite exceptional.

Another theme in the creativity–innovation model is to trace pathways by which highly creative and influential people and events have established or inspired "schools" and movements for the spread and adoption of new ideas in practice. Social and economic historians and philosophers of science are active in this field of enquiry (see Ville, this volume).

Technological (and Other) Innovation Sparks Creativity

The second pathway reverses the apparent "natural" order. Here innovation stimulates or enables creativity. But innovation–creativity is a natural part of a sequence that reads innovation–creativity–innovation. Nathan Rosenberg (1982) documents the impact of new technologies on generating new science. Jane Marceau (Chapter 3, this volume) designates technological change (and its rapid diffusion) as one of the core components of an innovation system. It would be difficult to conceive of any creative activity which does not relate to or utilise existing or changing technology. Beethoven's nine symphonies were built on the technology of musical instruments that comprise an orchestra; indeed, the orchestra itself is a cultural innovation.

West (Chapter 2, this volume) observes "rather than originating from science, new technologies often themselves precipitate new science, aimed at understanding more deeply what has been observed to work and improving it". Numerous examples are given in this volume of how new technology (innovation) leads to creativity in the specific or cognate domain. Consider the impact of nanotechnology and miniaturisation for the development of creative new procedures and techniques in microsurgery, or the impact of new computer technology on creativity in installation art and photography. Mark Dodgson (this volume) comments on the impact of new information technology and visual technologies on creative design, architecture and new products. The new technologies encourage "play", i.e. the creative part of the equation which comes between the initial idea and action (implementation). Jane Marceau (Chapter 3, this volume) comments on the role of innovation in IT as a platform for creativity (new product designs) and innovation in the textile and clothing industry (shorter and more efficient supply chains, Just-in-time manufacturing). Simon Ville (this volume) comments on how "macro"-inventions, e.g. the automobile and electricity, triggered a surge of creativity and innovation in other industries. Brian Fitzgerald (this volume) comments on the actual and potential impact of the Internet on creativity. In Chapter 16 (this volume) Mann comments on how new information and communication technologies are transforming the way research is conducted, from virtual teams, to completing surveys online, to statistical software packages for data mining and analysis. Peter Doherty (cited in Mann and Chan 2009) describes how new powerful statistical tools that examine large data sets have become a tool for identifying research problems and for generating fresh hypotheses.

There are variations in the innovation–creativity pathway in and across domains. Some pathways are direct (e.g. invention of the saxophone leads to a surge in jazz creativity) and others indirect (e.g. the invention of polymer house paint leads to new developments in pop art). A feature of the innovation–creativity pathway is that it is basically unplanned and quite unpredictable. But it is increasingly recognised that innovations from parallel and cognate disciplines can combine to produce creativity. New techniques and instruments (innovations) from several diverse fields can provide a vehicle for discovering, describing and measuring something new. An example is how new techniques and tests from neuropsychology (problem-solving and memory tests), brain anatomy (brain imaging) and physiology (single cell analysis) have combined to produce new insights into dementia and Alzheimer's disease.

Both creativity–innovation and innovation–creativity describe essentially linear pathways between a particular concept or idea and a particular discovery, invention or product. The designation of pathways as linear suggests order and the possibility for setting things right if something seems to be off track.

Creativity and Innovation Are Parts of a Dynamic System

A third model posits creativity–innovation within a complex, dynamic system of major components linked by multiple relationships (pathways). The components are the key participants, organisations and institutions that together make up the particular innovation system—national, regional, city, industry, etc. (see West and Marceau, Chapters 2 and 3, this volume). Creativity and innovation respond and build on each other in a series of connections and feedback loops within the system, with sources of creativity and innovation coming from the components separately and together.

Freeman (1995) maintains that the level of national innovation capacity (and presumably the level of creativity) is a function of interactions between a set of players and institutions from the research and non-research sectors. From an innovation systems perspective, technical innovations emerge through a complex process of combination and diffusion of different knowledge elements, involving "complicated feedback mechanisms and interactive relations involving science, technology, learning, production, policy, and demand" (Edqvist 1997, 3). Jonathan West (Chapter 2, this volume) points out:

> The relations between creativity in science and innovation are . . . complex. In some instances, new science gives birth to new technology and commercial innovation . . . Just as often, however, commercial innovation gives impetus to new science, or draws upon existing science in ways its originators had no way to foresee. Serendipity thus plays an irreducible role in the relations between science and innovation. Innovations often spring from applications of science that are quite unexpected by their original scientific discoverers.

A feature of industry innovation systems is a focus on the quantum of creativity and innovation in the total system (and how the two are related), not any particular innovation, although that is relevant. The same logic applies to analysis of creativity–innovation links in any institution, for example, universities as creators and discoverers of knowledge. Jane Marceau (this volume) presents an example of the system approach in her analysis of innovation at the high-tech end of the Australian building construction industry (large office buildings, stadiums, bridges, etc.). In essence the high-innovation profile and performance of the industry is due to the small, highly creative companies that supply new products, materials, design solutions and professional services to the large, lead companies that run the industry. The building construction case is an example of how the connection between creativity and innovation in an industry is a function of many players, e.g. lead and supplier companies, industry and government regulators, education and training institutions, availability of new technologies and firms willing to apply them.

The system perspective also draws our attention to industry networks of firms that share knowledge, cooperate and compete to stimulate creativity and innovation (Marceau, Chapter 3, this volume). Peter Hall (1998) proposed that a network of cotton spinners initiated the Industrial Revolution in Manchester and another network made Glasgow the centre of world shipbuilding. Innovation networks help explain how Berlin came to dominate the early electrical industry, Detroit the manufacture of automobiles and Tokyo-Kanagawa the production of electronics. Innovation clusters and networks also underpin the success stories of Silicon Valley in biotechnology and electronics and Milan in textiles, fashion and automobiles.

Creativity and Innovation Intersect in Great Cities and Places

The system model outlined in the preceding describes an industry innovation system based on business, science and technology. But note how Peter Hall (1998) located industry innovation networks within particular cities, Manchester, Glasgow, Berlin, Detroit and Tokyo. We posit another dynamic system model to describe the outpouring of creativity and innovation in particular places at particular times across domains, including technological, cultural and social innovation. The great cities of the world are prime examples of dynamic systems of creativity and innovation. They are places where creativity and innovation flourish across many domains. Peter Hall referred to "the unique creativity of great cities" (1998, 7). Jane Marceau (Chapter 4, this volume) discusses the impacts of city and place on creativity and innovation.

Cities are dynamic systems because of a common set of antecedent factors which include progressive institutions of governance; fine universities, libraries and concert halls; a community of highly educated and skilled residents; leading firms; people who come to the city with exciting new ideas; freedom to experiment with ideas; a rich social and cultural milieu; and so on. But they are dynamic systems because great cities hum with creativity and innovation across many domains, one domain inspiring and nourishing the other in a melting pot that generates new ideas and opportunities. The pathways under this model are more than the relationships between components of innovation that underpin industry in and near the city (see the preceding), but also the connections and links between the many domains—education, medicine, science, architecture, design, art, music, theatre, law, governance, social philosophy, etc.—in which creativity and innovation are expressed. This model is difficult to specify as a set of pathways because the city is a place of seemingly infinite possibilities for people with creative ideas to meet people prepared to back them. The coming together of "creatives", "innovators" and "entrepreneurs" in the city generates opportunities for blending ideas from many sources into new combinations.

Thus creativity and innovation literally intersect in the streets and meeting-places of great cities producing fusions of new ideas and possibilities across fields and domains. The creativity of the city is expressed even in the design and redesign of the city itself. Examples are Paris in the eighteenth century and now Shanghai in the twenty-first century. Hall noted that some cities are famous as centres of cultural and social innovation while others are famous as centres of technological and economic innovation. Some cities are famous for both. Los Angeles, for example, is renowned for its fusion of artistic creativity and technological innovation in the movies, television, communication and multimedia. How do we trace the connections between creativity and innovation in this configuration? The simple answer is with great difficulty. But we can look at change and transformation in great cities and look for differences across domains, for example, as waves of migrants with new skills bring new vitality to the city or new institutions are established.

Creative Work and Innovation Can Be Organised and Managed Creatively

Increasingly, creative work is performed in teams, organisations, networks, partnerships and collaborations. Such work can be creatively organised, managed and supported. The notion that the creativity–innovation connection can be enhanced by smart management echoes the title of Bennis and Biederman's (1997) book *Organizing Genius: The Secrets of Creative Collaboration*. The nature of research collaboration can itself be creative, for example, in planning how best to collaborate, who works together and how they will work together. The sub-discipline of management of technology and innovation explains how organisations frequently struggle to design their structures and procedures to foster creativity and innovation alongside requirements for centralisation, accountability and coping with uncertainty (West, Chapter 12, this volume).

Institutional innovation (i.e. the establishment of new institutional forms) can give rise to creative research and innovation. The classic example is establishment of university laboratories and industrial R&D labs in the late nineteenth century. The scientific laboratory was an institutional innovation that involved the systematic organisation and conduct of research in project teams to solve fundamental problems or in the case of industry to develop new products and processes (cf. Freeman 1995; Buderi 2000). Thomas Edison once boasted that his greatest invention was the industry lab itself, a view echoed by a leading physicist who once said that the greatest invention of the nineteenth century was the method of invention itself (Freeman 1995).

Creativity can also be applied to innovation policy and to human capital, business, law and governance practices. There is growing evidence that smart planning and intervention can make a difference in industry and enterprise innovation systems. West (Chapter 2, this volume) cites the case of innovation in the Taiwan semiconductor industry as an example of clever government design of an innovative industry from the ground up.

In sum, there are many conceivable pathways between creativity and innovation. I outlined five models in which creativity and innovation are connected, in some cases more directly and strongly than in others. The first and second models describe linear pathways between creativity and innovation. The third and fourth models describe dynamic systems in which creativity and innovation components interact (third model) and creativity and innovation domains intersect (fourth model). A fifth model suggests that the work of creativity and innovation itself can be creatively organised and managed. The different pathways help explain why creativity and innovation appear tightly coupled in some areas but weakly related in others. The challenge is to trace pathways and examine the factors that determine particular pathways.

STUDYING THE CONNECTIONS BETWEEN CREATIVITY AND INNOVATION

The final step in discussion of pathways between creativity and innovation is to mention several research methods from the social sciences that can be used for tracing and examining pathways. Invariably the research question will help determine the methods employed.

Case Studies

The case study is a well-known method for in-depth study of creativity and innovation in organisations and within industries. Several examples appear in this volume, including Dodgson's studies of creativity in design organisations and Marceau's studies of creativity and innovation in industry sectors such as textiles and clothing and building construction. West describes a case study of the Taiwan electronics industry. Case studies are useful for examining accounts from different perspectives. They are especially useful for formulating hypotheses and testing interpretations when the methodology involves paired comparison between two matched companies or industries. For example, Marceau compared differences in level of innovation between two segments of the building construction industry. The key difference was that lead companies in one sector were open to creative products, processes and services from small suppliers, while lead companies in the other sector were indifferent to new technology.

Field Studies and Experiments

Donald T. Campbell (1969) wrote a classic article "Reforms as Experiments" in which he discussed the methodology of planned policy reforms, such as introduction of a new tax incentive for R&D expenditure, as an opportunity to conduct a before–after comparison of innovation activity and performance. A control group in which the tax did not apply to a particular industry would provide a useful comparison especially if that industry was designated as a "waiting" control, i.e. it would subsequently receive the tax incentive. Campbell argued for careful baseline measurements in order to provide reliable evidence for the impact of the reform. To my knowledge there are few if any field experiments of creativity–innovation. There have been numerous surveys and experimental studies of the benefits of training programs and interventions for fostering creativity and innovation. Claims have been made about the efficacy of training programs based on these studies, but, as Sweller and Mann argue in this volume, there is little evidence that the programs (usually a tool-kit of techniques) have any real impact on actual creativity.

Surveys and Archival Analysis

Studies of national innovation systems by Michael Porter, Scott Stern, Jeffrey Furman and others involve the statistical analysis of economic data sets (from surveys, government reports and archives) for a sample of countries across industries and over time (e.g. Porter and Stern 2001; Furman, Porter and Stern 2002). This is a very useful approach for studying the correlates of innovation and for identifying innovation clusters. Extension of this approach to incorporate surrogate measures for creativity along with measures of innovation (such as patents) will assist macro-level analysis of the links between creativity and innovation.

Multilevel Statistical Analyses of Surveys and Data Sets

A promising approach for linking creativity and innovation is the use of multilevel statistical analysis to identify the effects of individual-, group-, organisation- and industry-level factors separately and in combination on creativity and innovation. An entire volume has been published on issues in multilevel analysis of creativity and innovation (Mumford, Hunter and Bedell-Avers 2008). Multilevel analysis of creativity will ordinarily encompass the individual, team and organisation as levels of analysis (e.g. Taggar 2002; Pirola-Merlo and Mann 2004). Multilevel analysis of innovation will ordinarily encompass nation, industry sector and institution (e.g. Furman, Porter and Stern 2002). A study by Lorenz and Lundvall (2010) is an example of the use of multilevel analysis to obtain a fuller picture of factors impacting on creative work across countries in the European Union. Using multilevel logistic modelling they analysed individual characteristics (education level and work experience), organisation of work (e.g. workplaces where work is organised to promote knowledge diversity) and national-level data (national education and training systems and labour markets) to explain differences between countries in participation in creative work.

Historical Analysis

Finally, historical analysis of archives, documents and evidence from different periods can assist in identifying the major influences and events associated with technological, social and cultural innovation and other transformations associated with flows of creativity. Jenny Uglow (2002) shows how the great eighteenth- and nineteenth-century thinkers Darwin, Watt, Wedgwood, Boulton, etc., combined creativity and innovation in their passion for chemistry, botany, engineering, business, politics and poetry. Dean Simonton (1996) has used socio-historical analysis to discuss genius and creativity in different periods. Simon Ville (this volume) uses economic historical analysis to examine the macro-level factors responsible for transformation in key industries.

In sum, we believe that advances in the study of creativity and innovation will follow from a clear definition of the core concepts, specifying the level of creativity or innovation under examination, identifying possible and actual pathways and using reliable research methods to examine multilevel linkages.

BIBLIOGRAPHY

Amabile, T. M. 1996. *Creativity in context: Update to the social psychology of creativity*. Boulder, CO: Westview Press.
Bennis, W., and P. Biederman. 1997. *Organizing genius: The secrets of creative collaboration*. Cambridge, MA: Perseus Books.
Berkun, S. 2007. *The myths of innovation*. Sebastopol, CA: O'Reilly Media.
Buderi, R. 2000. *Engines of tomorrow*. New York: Simon and Schuster.
Campbell, D. T. 1969. Reforms as experiments. *American Psychologist* 24 (4): 409–29.
Davila, T., M. Epstein, and R. Shelton. 2006. *Making innovation work*. Philadelphia: Wharton Business School Press.
Edqvist, C., ed. 1997. *Systems of innovation: Technologies, institutions and organizations*. London: Pinter Publishes/Cassell Academic.
Fabrizio, K. 2006. The use of university research in firm innovation. In *Open innovation: Researching a new paradigm*, ed. H. Chesbrough, W. Vanhaverbeke, and J. West, 134–160. Oxford: Oxford University Press.
Freeman, C. 1995. The "National System of Innovation" in historical perspective. *Cambridge Journal of Economics* 19: 5–24.
Freud, S. 1910. The origin and development of psychoanalysis. *American Journal of Psychology* 21: 181–218.
Furman, J. L., M. E. Porter and S. Stern. 2002. The determinants of national innovative capacity. *Research Policy* 31: 899–933.
Hall, P. 1998. *Cities in civilization*. New York: Pantheon Books.
Levitt, T. 1981. Ideas are useless unless used. *Inc. magazine* 3 February 1981: 96.
Lorenz, E., and B-A. Lundvall. 2010. Accounting for creativity in the European Union: A multi-level analysis of individual competence, labour market structure, and systems of education and training. *Cambridge Journal of Economics*. Advance Access published online on 21 April 2010 as doi:10.1093/cje/beq014.
Mann, L., and J. Chan. 2009. *Fostering creativity and innovation. Report of the 2008 Annual Symposium, 18–24, ASSA annual report*. Canberra: Academy of Social Sciences in Australia.
Morelock, M., and D. H. Feldman. 1999. "Prodigies". In *Encyclopaedia of creativity, vol. 2*, ed. M. Runco and S. Pritzker, 449–56. San Diego: Academic Press.
Mumford, M. D., S. T. Hunter and K. E. Bedell-Avers, ed. 2008. *Multi-level issues in creativity and innovation*. Oxford: JAI Press.

Pirola-Merlo, A. and L. Mann. 2004. The relationship between individual creativity and team creativity. Aggregating across people and time. *Journal of Organizational Behaviour* 25(2): 235–257.

Porter, M. E., and S. Stern. 2001. Innovation: Location matters. *MIT Sloan Management Review* 42 (4): 28–36.

Rosenberg, N. 1982. *Inside the black box*. Cambridge: Cambridge University Press.

Sawyer, R. K. 2006. *Explaining creativity: The science of human innovation*. New York: Oxford University Press.

Simonton D. 1996. *Greatness: Who makes history and why*. New York: Guilford Press.

Stokes, D. 1997. *Pasteur's quadrant*. Washington, DC: The Brookings Institution.

Taggar, S. 2002. Individual creativity and group ability to utilize individual creative resources: A multilevel model. *Academy of Management Journal* 45 (2): 315–330.

Uglow, J. 2002. *The lunar men*. New York: Farrar, Straus and Giroux.

Waldman D. A. 1994. "Transformational leaders in multifunctional teams". In *Improving organizational effectiveness through transformational leadership*, ed. B. Bass and B. Avolio, 84–103. Thousand Oaks, CA: Sage

Zucker, L. G, M. R. Darby and M. B. Brewer. 1998. Intellectual human capital and the birth of US biotechnology enterprises. *American Economic Review* 88 (1): 290–306.

16 Creativity and Innovation
Principles and Policy Implications

Leon Mann

In the introductory chapter we described the burgeoning interest in creativity and innovation, the concept of the creative knowledge economy and the interest of governments and other institutions in fostering innovation. In this final chapter we do three things: first, revisit the main themes in the book; second, summarise ideas relevant to policy implications and recommendations; third, offer a message about understanding the sometimes unforeseen impacts of technological and social innovation (see also Mann, Chapter 15, this volume, on other ideas relevant to policy).

WHAT THE SOCIAL SCIENCES SAY ABOUT CREATIVITY AND INNOVATION

From social science perspectives, we examined a matrix of overlapping systems, institutions, places and domains where creativity and innovation occur; these encompass nations, regions, cities and precincts, business firms and public research organisations and knowledge communities, and involve diverse fields such as science and technology, design and the arts. The chapters have discussed components of national innovation systems (West, Gans, Marceau), cities as places of creativity and innovation (Marceau); innovation in a range of industry sectors, such as textiles and clothing, building construction, design and new media (Marceau, Ville, Fitzgerald, Dodgson); factors driving and impeding creativity and innovation in R&D organisations and firms (Dodgson, West, Mann); factors in research team creativity and performance (Mann); the significance of knowledge communities in determining and recognising creativity (Chan); the purpose and impacts of legal instruments for protection of intellectual property as incentives for or impediments against innovation activity and creativity (Christie; Fitzgerald); the role of schools in supporting creativity (McWilliam); and the "architecture" of creative ability in the information storage and processing capability of the human mind (Sweller and Mann).

A number of observations can be made about the contributions of social sciences to the study of creativity and innovation. First, there is no uniform social science approach to creativity and innovation, rather a range of disciplinary perspectives. The disciplines have a common interest in examining creativity and innovation as activities performed by individuals and increasingly by groups, organisations and communities of practice across a wide range of settings and knowledge domains. Second, various social science disciplines tend to focus on *either* creativity *or* innovation in accordance with their core subject matter. Thus psychology and education tend to study creativity as they are interested in individual abilities and behaviours; social psychology and sociology tend to examine both creativity and innovation as they are concerned with groups, organisations and communities involved in creative activities for innovation or from which innovation follows; whereas economics, history, law, political science, policy and management studies tend to be interested in innovation as they deal with institutions and systems. Third, locating creativity in a social/cultural/historical/evolutionary context moves the focus from individual creators and their abilities and characteristics to the social conditions that support, inhibit, constrain or enable creative work to take place (cf. Feldman and Benjamin 2006). Fourth, the social science disciplines operate at different levels of analysis and can connect across different levels—encompassing individual, group and team, organisation, community, city and nation. Accordingly, there is opportunity to use multilevel analysis to examine connections between creativity and innovation (see Lorenz and Lundvall 2010). Fifth, the study of creativity and innovation has been enlarged by new and emergent fields, disciplines and perspectives. Marceau (Chapter 4, this volume) draws on the new discipline of economic geography to examine cities as places of innovation. Dodgson (this volume) examines innovation from the discipline of management. The subfield of innovation policy studies draws on several social science disciplines. Entrepreneurship has emerged as a new related field. Most chapter writers in this volume acknowledge how their analysis is based on their "home" disciplines and also other cognate disciplines. History and philosophy of science, culture studies, media studies, knowledge management and information sciences, engineering and design studies are also significant contributors.

MAIN THEMES IN THIS VOLUME

There are seventeen recurrent themes in this volume:

1. Creativity and innovation are central to thinking and planning for national change and transformation, productivity and performance and social and economic success. This is evident in government interest and policy (Australia has a Minister for Innovation, Industry Science and Research), OECD policy research and government expert reviews. Dodgson, West, Marceau, Mann and Gans, chapter writers in this volume, have contributed to Australian government reviews on innovation priorities, policy and performance.

2. It is possible to foster creativity and manage and improve innovation across settings and domains, and specifically in business firms and research organisations.

3. Nations differ in innovation systems and performance. The evidence suggests components in successful systems in Western countries are public research expenditure, availability of well-trained human capital, intellectual property protection and free trade. The hand of government is clear and visible. To understand why some nations are more innovative, map the major components in each system and examine the "fit" and connections between components (West, Marceau). Other relevant factors are trust, social capital and open communication between organisations (Lundvall et al. 2002).

4. Innovation systems are connected at different levels, for example, between nation and industry, between nation, region and city, to drive innovation capacity and performance at each level (Marceau, West).

5. Industries differ in innovation systems and what determines success, e.g. in manufacturing, pharmaceuticals, electronics, building construction, textiles and clothing. The factors include proximity to other firms, product supply chains and adoption of new technology (Marceau, West, Ville). Industrial innovation drives the innovation performance of a nation, region and city. Thus Finland performs well because its industries produce innovative design, mobile phones and telecommunications.

6. Cities, as innovation systems, connect with other systems of innovation from which they draw direction and opportunities and to which they contribute ideas, inspiration and energy. Factors associated with creativity and innovation in cities include good governance and incentives to attract and retain firms, and proximity to fine universities and research institutions (Marceau).

7. The pulse of creativity and innovation fluctuates across historical epochs and eras. Factors associated include migration, breakthroughs in platform technology in another domain, e.g. printing, transport, energy, communication technologies and government policies, and incentives (Ville, Gans, Christie, Fitzgerald).

8. Points 3 through 7 lead to the conclusion that a "one-size-for-all" formula does not work for the design of innovation systems for nations, cities and regions, industry sectors and domains. Each system must be examined for distinctive features—history, cultural values, knowledge flows, relationships between key players before offering solutions (West, Marceau, Gans).

9. Creativity and innovation are social activities and are increasingly sponsored, organised and managed; this holds for both arts and science. Clearly,

some individuals are more exceptionally creative than others, but it is a myth that creativity is the work of the lone genius.

10. Further to point 9, creativity and innovation are located in social structures and in teams, organisations, communities of practice, and social networks which provide ideas, information, tacit knowledge, feedback and expertise from other disciplines to support, encourage and validate creative and innovative activity (Chan, West, Dodgson, Mann). Scandinavian scholars (Lundvall, Edqvist) emphasise trust, communication channels and collaborations as significant factors.

11. Creativity is fundamentally domain specific: creative people are very rarely creative across a range of fields and domains. They are bright, motivated people who are creative through immersion and deep knowledge in a particular domain. This highlights the importance of in-depth knowledge in a domain to be creative. A corollary is that the knowledge and talent necessary for creative work in one domain may be different from what is required in another.

12. It follows from point 11 that attempts to "teach" creativity as a set of generic rules and principles ("think outside the box") and techniques (e.g. "do some brainstorming") is irrelevant for genuine creativity but may be useful as a tool for "everyday creativity", i.e. routine problem-solving (Sweller and Mann, McWilliam).

13. Creativity and innovation can, up to a point, be managed. "Creatives" are motivated by the intrinsic stimulation of their activity and by freedom to explore ideas and try new things (cf. Amabile 1996). Innovation is more amenable to management, *e.g.* through the design of organisational structures and processes for information flow and exchange, tacit knowledge, creative "play" with ideas, team-work, collaboration, knowledge networks and knowledge transfer (Chan, Mann, West, Dodgson).

14. Creativity and innovation are nourished by "open" innovation systems (cf. Chesbrough, Vanhaverbeke and West 2006). In an open innovation system the organisation establishes purposive inflows and outflows of knowledge to build innovative capacity. Ideas are gathered from many sources outside the organisation.

15. There are connections between innovation in technology, culture and society. Marceau argues for citizen participation in new forms of governance to find solutions for cities (an example of social innovation). Fitzgerald makes a case for relaxation of copyright laws and entitlement to reuse of copyright material for cultural and social innovation. The two examples refer to the connection between technological, social and cultural innovation.

16. Government and public-sector institutions have a compelling role to devise the legal and IP regimes, opportunities and incentives to buttress creativity and innovation. Government has a role to fund innovation activity in areas of "market failure" (perhaps more precisely "market non-interest"); public research organisations have a role to perform fundamental research of national and international significance.

17. Firms dependent on competitive new products and services struggle with the challenge of how best to design their organisational structures and culture to support creative and innovative activity. Universities and schools coping with limited budgets are stretched to prepare the skilled and literate workforce capable of developing new technology and innovation (West, McWilliam, and all chapter writers).

POLICY IMPLICATIONS

Policy analysis relating to creativity and innovation involves identifying policy options, implementation and assessment (Potts 2009). The field of innovation policy has its own journals, e.g. *Innovation: Management, Policy & Practice* and *International Journal of Foresight and Innovation Policy*. However, comparatively little is available on creativity policy. In this section we suggest policy recommendations and implications based on work in this volume. We are interested in principles for understanding and working with creativity and innovation systems and formulation of recommendations for fostering creativity and innovation for particular institutions and players. We bring the discussion together under ten headings: governments and national innovation systems; industry innovation; location and place; technology; universities; intellectual property; firms and organisations; field and habitus; schools and education; and individual capability.

Governments and National Innovation Systems

The analysis of national innovation systems (Freeman 1987; Lundvall et al. 2002) is the starting point for understanding the institutions, structures and practices that underpin national innovative capabilities and what must be done to improve performance. Fundamental to the national innovation system are technical and scientific institutions, the national education system, industrial relations, government policies, cultural traditions and other national institutions (Freeman 1995, 5). Readers interested in different national innovation systems will find useful the *OECD Reviews of Innovation Policy*, which covered Switzerland, Luxembourg, New Zealand, South Africa, Chile, Norway, China, Hungary, Korea and Mexico in the period 2006–2009. The components identified in these reviews include IP protection, R&D activity/and expenditure and higher education participation. In this volume we do not look at global innovation systems, e.g. international R&D consortia, migration of skilled knowledge workers and international collaborations, a feature of many high-tech industries.

In their chapters, Jonathan West, Jane Marceau, Joshua Gans and Simon Ville make recommendation about national innovation systems. West's key recommendation is that policymakers interested in reconfiguring their national innovation system must recognise there is no single, uniform strategy. As Cinderella's ambitious sisters discovered to their dismay, one size does not fit all. But he cites evidence based on comparative studies of national innovation systems (Furman, Porter and Stern 2002) to identify several factors that matter for greater innovation intensity (higher level of patenting) and R&D productivity: the commitment of government resources to education and research, IP protection (patent and copyright laws) and openness to international trade. Other factors are a higher proportion of research performed by the higher education sector, the degree of technological specialisation and the country's knowledge "stock". Each of the OECD countries that increased innovative capacity in the period 1973–1996—Japan, Sweden, Finland, Germany—implemented policies that encouraged investment in science and engineering (e.g. by establishing and supporting technical universities) and also stimulated competition on the basis of innovation (e.g. through adoption of R&D tax credits and opening of markets). The case of the Taiwan semiconductor industry demonstrates how astute government policy can improve the innovation and productivity of a nation.

West makes three policy recommendations. First, government should complement, not attempt to substitute for or compete with, the activities of private companies. Government should aim to strengthen the market position and capabilities of private firms. Government policy should "do no harm": it should not distort market incentives by attempting to substitute public-sector activity for the private sector. It should limit its involvement to fields in which infrastructure, capability and resources are demonstrably required to support innovation and it is not feasible for private companies or markets to become involved (see also Gans, this volume, in regard to private-sector "market failure" for major innovation). Second, government policy should be directed only towards specialised industry sectors of sufficient size, weight and potential to matter to the national economy. This implies that innovation policy should go beyond "research-intensive" industries such as ICT, biotechnology and nanotechnology. Third, governments should ensure the long-term strength of their "heavy-weight" industries and strategically plan where to locate supporting facilities, institutions and resources. West and Marceau (Chapters 2 and 3, this volume) do not identify the *most* important components in a national innovation system. They suggest all components are vital. But evidence suggests that investment in public-sector research (cf. Gans, this volume), a well-educated and skilled workforce and the free flow of ideas and trade are absolutely essential to national innovation.

Joshua Gans (this volume) argues the case for government support to deal with the problem of "market failure", i.e. private-sector reluctance to invest in risky, expensive projects that are vital for national interest. He argues for balance and complementarity between public and private research systems (see also West). He advocates IP protection, tax benefits and subsidies, competitive reforms and grants as mechanisms for governments to provide incentives for innovation. He also recommends government prizes (e.g. the first to find a cheap source of electricity), research tournaments, matching grants and government advanced purchase commitments, e.g. for battery-operated cars, to stimulate product innovation.

An important caveat in policy formulation for national innovation systems is to recognise that the research evidence in this volume draws heavily on Western market economics analysis and on high-tech innovation. The applicability of the analysis to non-Western economies must be questioned.

Simon Ville, an economic historian, maintains that history tells us to value creativity and unorthodox thinking, invest in human capital, attend to the learning system and how we learn, tolerate unorthodoxy, support fundamental undirected research, accept that many areas of creative thinking and research will not produce a tangible outcome, foster a culture of trust and cooperation—the concept of social capital (cf. Lundvall 2002 and Edqvist 1997), treat science as a public good and foster interaction between scientific researchers and practitioners.

Industry-Sector Innovation

Jane Marceau (Chapter 3, this volume) deals with industry innovation systems within the overall national system. Different drivers stimulate innovation across industry and service sectors, such as textiles and clothing, farming, transport, health care, manufactures, building construction, pharmaceuticals and higher education. As with national innovation systems, "one size does not fit all". Therefore, different policies are required to make particular industries more innovative. Marceau examines two dimensions that affect the innovation performance of an industry: information sharing and the exercise of power and influence between key players in the industry. Her analysis leads to two recommendations. First, that strong and approximately equal flows of information and knowledge exchange between the main players, viz. producers, users, R&D providers, training institutions and regulators, is associated with greater innovation. Second, that more or less equal exercise of power by the main players can boost innovation, while excessive use of power stifles innovation. However, there are exceptions, for example, when a dominant player takes charge during an industry down-turn or crisis. Adoption of new technology can lead to a shift in power and direction setting in an industry. Accordingly, there are policy lessons for all industry players to be receptive to opportunities opened by new technologies.

History teaches lessons about new technology as an enabler of innovation across other industries. According to Simon Ville (this volume), during the nineteenth century "macro"-inventions such as electricity and railways had a transformational effect on a wide range of industries and created new ones. Ville maintains that governments and industry must envisage the potential effects of new foreign-derived technologies, facilitate their transfer and adaptation and establish incentives for their uptake as they will impact many industries. In many high-growth industries there are benefits from being an adapter and user-nation rather than a leader. Information technology is an example. With globalisation and deregulation small economies are able to compete at the sub-industry level.

Location and Place, Cities and Regions

Location is the setting and vehicle for innovation (Marceau, Chapter 4, this volume). There are different innovation profiles for firms and industries across geographic areas and regions. Proximity to other firms (industry clusters), institutions of learning (for new knowledge), markets and customers (for local knowledge) are important for firms involved in innovation, especially product innovation. Industry clusters and networks encourage collaboration, productive competition and market development. Zucker, Darby and Brewer (1998) found that firms located close to good universities benefit from a "spillover" of knowledge. Porter and Stern (2001), employing OECD data, showed that the location of a firm's R&D facilities (usually in or near to cities) is very important for technology innovation. Regions also matter for innovative activity. Famous examples include Silicon Valley in California for electronics and biotechnology and Emilio Romagna in Italy for automotive industry and food products, which draw on the creative energy of adjoining cities.

Major cities are innovation hubs. Cities are where new ideas, research programs, technology development and new business products and processes happen. They are places where skilled people performing creative work and leading firms with global reach are located. Cities are where government and policymakers converge and where universities and research institutes, corporate R&D labs, science-technology parks and precincts cluster. But this is an ideal picture. Many cities are in decline. A policy implication is that government must ensure the vitality of key cities as powerhouses of innovation. Marceau writes, "The transformation of spaces into innovative places is perhaps the most complex endeavour that policymakers and business people alike have ever had to deal with". She recommends that cities must attend to governance arrangements for infrastructure development, urban change and renewal and incentives to experiment with new policies.

The OECD (2006) review of problems of urban decay and transformation in major European cities identified dilemmas for governments. The dilemmas include reconciling the interests of national, regional and city governments in accommodating civil society goals and ideals, e.g. social inclusion, affordable housing, access to cheap transport and essential services, with the needs of industry and business in the city. Reconciling public and private interest and building public–private partnership is a top priority and requires institutional and organisational creativity. The OECD (2006) review also examined issues of coordination, for example, in building cooperative networks between major cities and regions, e.g. by twinning universities and regions and locating different parts of major technology projects across several places. Packages of policies, not single-issue interventions, are recommended to address urban sustainability and innovation. Marceau maintains that technological innovation can assist social innovation in city governance. She argues that citizen's interest groups should be involved in governance along with business firms to devise solutions for the sustainability of cities and produce ideas for new institutions and projects. Citizens familiar with new IT and communications technology can participate in policy debates and contribute to place-based innovation.

Technology

Technology is a core component of all innovation systems. Technology transforms every domain of human knowledge and endeavour and is the platform for even greater innovation. The new information and communication technologies: enable information storage (West) and information exchange (Marceau); create opportunities for astute players to become dominant in innovation systems; open the way for small, smart "creatives" to adapt and find innovation niches; and change the way science and R&D are conducted. New IT affects quantitative research through data storage and retrieval, fast and sophisticated data analysis, and data "mining" (sifting through mountains of data to detect valuable trends and connections and to identify gaps). Nobel Prize–winner Peter Doherty describes data mining as a source of ideas for new hypotheses. The click of a computer mouse provides access to current and even advance publication of new knowledge. "Groupware" enables the efficient operation of "virtual" research teams working across several countries. In another domain, new communication technology has created social media (e.g. blogging, Twitter, YouTube, Plaxo, Facebook) to share new ideas and push innovations.

There are several policy recommendations for technology innovation. They include governments to revitalise technology-backward industries; ensure people understand how to use new technology; provide incentives to industry to make technology-friendly versions of products and processes for people who have language, health and education problems; and bundle new technology into user-friendly packages that include support services.

Universities and Higher Education

In Furman, Porter and Stern's (2002) survey of innovative capacity in seventeen OECD countries, they identified commitment of government resources to education and research and a higher proportion of R&D performed by the education sector as major factors predicting innovation intensity (higher patenting activity). Universities are major institutional players on the creative side of the creativity–innovation equation because of their expert faculty who perform fundamental research for new knowledge, talented research students and leading research centres and laboratories. But universities as sources of innovation should not be overstated. The highly innovative textile industry of Milan and Melbourne is driven by deeply experienced designers and manufacturers, not university professors. Universities are now taking an interest in the innovation–entrepreneurship side of the equation because they are under-funded by government. Accordingly, universities seek to commercialise their research to help cover their core teaching operations. Fabrizio (2006) notes that university research is increasingly associated with university control of IP, patent activity and commercialisation. The prospect arises that firms in the private sector will have limited access to public science to the detriment of industry R&D and innovation. The policy implications for government are obvious.

Gans (this volume) argues for retention of the traditional role of universities in the innovation system as the major producers of new ideas and concepts and accumulators of knowledge. He recommends government not to pressure universities to conduct research for short-term goals and commercial benefits, to the detriment of their traditional role in highly novel fundamental research.

Mann (2006) discusses the "collaboration imperative" between universities and industry for national innovation. Productive collaboration between universities and industry is a feature of a successful innovation system. But research collaboration is seldom easy because universities and industry hold different objectives and values. Partners to collaboration must have something to offer for collaboration to work. A strong adequately resourced university sector is in the national interest, and many observers have made that recommendation.

Mann (Chapter 10, this volume) argues that universities must do more to prepare research students for careers as creative researchers in industry and the public sector. The areas for research training are team leadership, project management skills, communication and negotiation, working effectively in teams, collaboration skills and multidisciplinary approaches and perspectives.

Universities should be exemplars of creativity and innovation in how they function in addition to the research they perform. Unfortunately, this is not always the case, and not only because universities (some founded in the twelfth century) are bastions of tradition. Glyn Davis, who heads a major Australian university, argues that it is hard for universities to be highly innovative because government funding models tend to discourage institutional diversity and change (Mann and Chan 2009).

Intellectual Property and Its Use and Protection

Jane Marceau (Chapter 3, this volume) identified intellectual property and its protection as a key component of a national innovation system. Furman,

Porter and Stern (2002), in their study of R&D innovative capacity in seventeen OECD countries from 1973 to 1996, observed that government policy on IP protection is a factor that really matters for innovation intensity (as indicated by number of patents). The perceived strength of IP protection, measured by survey responses, was significantly associated with innovative capacity. We have pointed out the limitations of the OECD studies for policy application in non-Western countries. However, there is considerable interest in what policymakers should do in regard to IP protection to encourage creativity and recognise and reward innovation.

Andrew Christie and Brian Fitzgerald are scholars in the field of intellectual property law. Christie (this volume) points out that patent law provides monopoly rights to inventors of new products processes and designs. The grant of a monopoly right to an inventor is a valuable entitlement because it can be exercised against people who copy the invention as well as those who later independently come up with the same idea. Christie maintains it is justifiable to afford this special protection to inventors because they bring something new to society. A policy recommendation that follows Christie's analysis is government should maintain strong patent laws to provide incentives and recognition for those who produce new inventions. Furman, Porter and Stern (2002) point out that IP protection is especially relevant for innovation in the pharmaceutical industry.

Fitzgerald (this volume) points out that copyright law provides rights to creators of literary and artistic works including music, painting, software code and architectural plans. He calls for relaxation of copyright laws to enable greater sharing of information and participation in creative activities. He argues that public policy must be responsive to changes in new technology that enable the free exchange of ideas that underpin and support creativity. He identifies the Internet in particular as a promoter of innovation and argues for open source innovation, i.e. free use of the Internet to organise activities, distribute production and create software. Fitzgerald makes three recommendations regarding the application of copyright law in the new information technologies. First, government should endorse the principle of "network neutrality", i.e. that all Internet traffic has equal access and speed; commercial users should not be favoured. Second, government should relax patent laws and copyright laws (especially) to provide greater opportunity for experimentation with new ideas and implementation of new, disruptive technologies. Finally, government should support the use and reuse of existing knowledge, e.g. through creative commons and open access to public-sector information. He also recommends a "peer to patent" system to provide greater efficiency in the patent system.

IP protection and regulation by governments worldwide has become a hotly contested and vexed issue. Policy in the area of IP is of interest (and of concern) to governments, the corporate sector, inventors and scientists, writers and citizens, as it has the most serious consequence for what is privileged monopoly over knowledge and technology and what is freely available. The proponents of strong IP protection argue it is essential for innovation and productivity. The most innovative economies are those with strong IP protection. Economies with weak IP protections tend to be less innovative and less competitive in the global economy. The opponents of strong IP protection argue for relaxation of tough legal regulations on IP and insist that IP protection is discriminatory and a threat to poor countries, and impedes the opportunity for everyone to use and benefit from new technology and knowledge. As editors we do not take a stance on the issue, except to recommend that governments in developed countries help developing countries to build innovation systems that will reduce their economic dependency.

Organisations: Ideas, Information and Innovation

In their chapters, Dodgson, Marceau, West and Mann discuss the creative and innovative activity of organisations. Dodgson suggests the concept of creative "play" as a link between creativity and innovation in organisations. The new design and virtual reality technologies (innovations) provide a set of new tools ("toys"?) for creative "play", which in turn provides platforms for creativity and innovation. "Play" includes activities where people explore, use templates and models, develop prototypes, rehearse and tinker with new ideas, often in a team working in a stimulating environment. Examples of organisations that produce innovations based on creative "play" include IDEO, Procter & Gamble and Gehry Technologies (architect Frank Gehry).

Mann (Chapter 10) recommends that research organisations especially should monitor how seriously they support their project teams to achieve innovation. A starting point is for organisations to attend to three aspects of support for innovation: Encouragement of innovation—does the organisation value and encourage innovation? Resources—do teams have adequate facilities, materials, information, time and access to expertise? Empowerment—do team members have the autonomy to develop new, risky ideas?

Organisations benefit from "open innovation" systems, i.e. porous boundaries through which ideas and information from outside the organisation are readily sourced and accessed (Chesbrough, Vanhaverbeke and West 2006). West (Chapter 12, this volume) argues that the existential dilemma of all large modern organisations is how to cope with the problems of too much information, complexity and uncertainty and remain innovative. West is pessimistic because neither a centralised organisational structure nor a decentralised structure can cope with this burgeoning problem. The imperative to be competitive or perish drives the search for solutions to the dilemma, in itself a creative response. We conclude that small organisations in niche areas that are adept at use of information technology for handling information and that apply communication technology for coordination will survive and be innovative. But West has issued a warning about information overload in organisations. This is a prime area for policy development and recommendations, one of which is to learn from what clever organisations are doing about the problem. Again, new technology might help. Companies such as Laing O'Rourke and IBM are providing innovation-related technologies capable of revolutionising the gathering, sharing and storage of information (Dodgson, Gann and Salter 2008).

Field, Habitus and Teams

Janet Chan (this volume) highlights the importance of the social dimensions of creativity. She points out it is useful to understand creativity in terms of different *fields* because each *field* has its rules, standards, power relations, access to resources and ways of inducting and socialising the potential creatives in the *field*. Thus, when considering creativity it is appropriate to think in terms of specific *fields* of creative practice, e.g. fine arts, literature, design, science and technology. Again, the dictum is: one size does not fit all. Through socialisation into a *field* or sub*field* the potential creative practitioner acquires *habitus* (the knowledge, skills or "feel of the game", modes of practice) of that specific *field*. Creative practitioners occupy different social positions in the *field* according to type and amount of "capital" they accumulate. In some *fields*, e.g. fine arts and basic science, the degree of autonomy is valued as capital, while in others, e.g. engineering and design, the level of expertise is capital. Capital may include access to resources, support and contacts that help generate further creative practice. Creativity is more likely when habitus and *field* are mutually reinforcing. Creative breakthroughs may occur when practitioners overcome obstacles resulting from a change of environment or domain, work together in multidisciplinary teams or are in a marginal position. There are several policy implications that flow from this analysis. First, one must be immersed in the knowledge of the *field*, including the formal and tacit knowledge. Second, one must learn the practices, power relations, where to get resources, whom to know, what the standards are for being acknowledged as creative and so on. Three, one must be prepared to step out of the comfort zone (try a new sub*field*, something different) and be willing to work with and be challenged by "creatives" from other *fields*. Chan also suggests several questions for research investigation: e.g. how is creativity understood and practised in different *fields*? What impact do changes in the *field* have on the *habitus* of creative practitioners and the judgment criteria used by creative communities?

Schools and Education

Training and education for useful knowledge and skills are a core component of all innovation systems. In this volume we view education more broadly than ensuring a highly skilled and well-trained workforce for industry. Erica McWilliam (this volume) deplores mediocre schooling in which "low threat, low challenge" is the norm. She wants school education to move closer to the "low threat, high challenge" learning culture that is important for building creative capacity in children. The creative capacity encompasses openness to experience, encouragement of novel ideas and thinking, opportunities to explore and build talent and so on. She recommends that policymakers, educational leaders and teachers attend to the following:

- Ensure that creativity goes beyond arts education in the curriculum and is also directed to bridging boundaries between disciplines for solving complex problems.
- Ensure that creativity (as both epistemological agility and artistic expression) is fostered from the early years to doctoral studies by ensuring creativity infuses the disciplines of sciences, mathematics and language learning, and not only the arts.
- Provide space in the curriculum for students to occasionally work in "design mode", i.e. to think, as many designers do, outside their cultural logics to find creative ways of engaging with familiar questions and dilemmas.
- Build collaborative teaching and learning cultures to foster creative capacity at all levels of education, and align curriculum pedagogy and assessment toward that goal.
- Build a creativity/innovation nexus through closer alliance with employing bodies.
- Undertake empirical research into the nature and value of creative capacity for learning and careers.

In Britain there has long been interest in encouraging creativity in schools (Craft 2005). As we write this chapter (in 2010), the draft Australian national curriculum for schools includes creativity in a list of ten general capabilities together with literacy, numeracy, thinking skills, self-management, team-work, social competence, ethical behaviour and intercultural understanding.

The Individual—Capacity and Capability

John Sweller and Leon Mann (this volume) discuss the psychology of creativity and its educational consequences, laying out the principles of a "cognitive architecture" based on memory, information-processing capacity and experience that underpins creative performance. They consider the "random generate and test" principle as "the ultimate source of all human creativity". They argue that knowledge specific to a particular domain is required for creative solutions to problems in that domain and that knowledge can be acquired only through a protracted and arduous process. In their words, "knowledge is the only available road to creativity ... and it is knowledge that forms the teachable aspect of creativity". The key recommendation is encourage capable and motivated people to immerse themselves in their field and acquire the vast amount of information and experience necessary for genuine creativity. The corollary recommendation is not to be deluded that you will become creative if you take a one-day training program in creativity-building skills.

CONCLUSION

The volume has covered the social sciences contribution to an understanding of creativity and innovation. But social science makes several other contributions to creativity and innovation. Mann (2001), in "Social Sciences and the Knowledge Economy", lists the following contributions: exploration of how knowledge is generated (a major theme in this volume); assisting in the design, use and adaptation of new technologies; understanding the impact of new technology and the new economy; policy analysis, advice and evaluation (touched on in this chapter); and understanding society and what makes us human.

An important task for social science is to understand the social impacts of new technology (Mann and Chan 2009). The application of new ideas and technologies can raise legitimate ethical and social concerns (consider abuse of the Internet) and cause individual and social upheaval. Not all innovations are welcomed and some are resisted and opposed. Social scientists have a role in understanding how people, especially the most vulnerable, are impacted by disruptive innovation. Social scientists, as all scholars, must maintain the highest standards of truth and transparency in discussion and debates about the risks and potential effects of new technologies, such as genetically modified crops, nanotechnology, gene technology and so on. Social scientists have a role to assist the most bewildered and vulnerable to comprehend and cope with social and economic dislocation caused by technological innovation, for example, in changes to access and use of services and amenities, in transport systems, health care, communication systems and in changes to required work skills, employment conditions and even unemployment.

The book title *Creativity and Innovation in Business and Beyond* reminds us that creative work ignites new ideas, concepts and actions across all spheres. Advances in knowledge can be used for good and sometimes for harm. At least one of the scientists who worked on the Manhattan Project, J. Robert Oppenheim, later regretted the bitter fruits of his team's creativity. Areas in need of greater creativity for positive purposes are health care, school and education; disadvantaged communities; and the rehabilitation and revival of cities, towns and neighbourhoods. There is also a vital need for the most creative thinkers to work on the urgent problems of climate change and renewable energy, protection and rehabilitation of natural resources, and the compelling task of peacemaking and conflict resolution. In this sense, the challenge of creativity and innovation goes beyond business and beyond the social sciences.

BIBLIOGRAPHY

Amabile, T. M. (1996). *Creativity in context: Update to the social psychology of creativity.* Boulder, CO: Westview Press.

Chesbrough, H., W. Vanhaverbeke and J. West, eds. 2006. *Open innovation: Researching a new paradigm.* Oxford: Oxford University Press.

Craft, A. 2005. *Creativity in schools: Tensions and dilemmas.* London: Routledge.

Dodgson, M., D. Gann and D. Salter. 2008. *The management of technological innovation.* Oxford: Oxford University Press.

Edqvist, C. 1997. Systems of innovation approaches—Their emergence and characterstics. In *Systems of innovation: Technologies, institutions and organizations*, ed. C. Edqvist, 1–35. London: Pinter Publishers/Cassell Academic.

Fabrizio, K. 2006. "The use of university research in firm innovation". In *Open innovation: Researching a new paradigm*, ed. H. Chesbrough, W. Vanhaverbeke and J. West, 134–60. Oxford: Oxford University Press.

Feldman D. H., and A. C. Benjamin. 2006. Creativity and education: An American retrospective. *Cambridge Journal of Education* 36 (3): 319–36.

Freeman, C. 1987. *Technology policy and economic performance: Lessons from Japan.* London: Pinter.

———. 1995. The "National System of Innovation" in historical perspective. *Cambridge Journal of Economics* 19: 5–24.

Furman, J. L, M. E. Porter and S. Stern. 2002. The determinants of national innovative capacity. *Research Policy* 31: 899–933.

Lorenz, E. and B-Å. Lundvall. 2010. Accounting for creativity in the European Union: A multi-level analysis of indvidual competence, labour market structure, and systems of education and training. *Cambridge Journal of Economics.* Advance Access published on 21 April 2010.

Lundvall, B-Å. 2002. *Innovation, growth and social cohesion: The Danish model.* London: Edward Elgar Publishing.

Lundvall, B-Å., B. Johnson, E. Andersen and B. Dalum. 2002. National systems of production, innovation and competence building. *Research Policy* 31 (2): 213–31.

Mann, L. 2001. Social sciences and the knowledge economy. *Dialogue* 20: 1–4

———. 2006. "Strength and length of partnership as key factor in research collaboration between universities and industry". In *How organizations connect: Investing in communication*, ed. G. Boyce, S. Macintyre and S. Ville, 74–99. Melbourne: Melbourne University Press.

Mann, L., and Chan J. 2009. *Fostering creativity and innovation. Report of the 2008 Annual Symposium, 18–24, ASSA annual report.* Canberra: Academy of Social Sciences in Australia.

Organisation for Economic Co-operation and Development. 2006. *Competitive cities in the global economy.* Paris: OECD.

Porter M. E., and S. Stern. 2001. Innovation: Location matters. *MIT Sloan Management Review* 42 (4): 28–36.

Potts, J. 2009. Introduction. Creative industries and industry policy. *Innovation: Management, policy and practice.* 11(2) 138–147.

Zucker, L., M. Darby, and M. Brewer. 1998. Intellectual human capital and the birth of the U. S. biotechnology enterprises. *American Economic Review* 88 (1): 290–306.

Contributors

Janet Chan is Professor at the University of New South Wales (UNSW) and a Fellow of the Academy of Social Sciences in Australia. She holds a BSc in Applied Mathematics, an MSc in Applied Statistics and an MA in Criminology from the University of Toronto. She moved to Australia in 1984 and completed her PhD in Law at the University of Sydney. Janet's research interests include reform and innovation in criminal justice, organisational culture and the sociology of creativity. She has worked as a consultant for a number of government organisations, was awarded nine Australian Research Council grants and has published numerous books and articles on criminal justice and policing issues, including *Changing Police Culture* (1997), *Fair Cop: Learning the Art of Policing* (2003) and *Reshaping Juvenile Justice* (2005). She is a practising visual artist and holds an MFA from the College of Fine Arts, UNSW. Her recent research projects have focused on understanding individual and collective creative practice among artists, scientists and art–science collaborations.

Andrew Christie is the foundation holder of the Davies Collison Cave Chair of Intellectual Property at the Melbourne Law School. He has bachelor's degrees in Science and in Law from the University of Melbourne, an LLM with distinction from the University of London and a prize-winning PhD from the University of Cambridge. Admitted to legal practice in both Australia and the United Kingdom, Andrew worked for many years in the intellectual property departments of major law firms in Melbourne and London. A former Fulbright Senior Scholar, Andrew has held appointments at the University of Cambridge, Duke University and the University of Toronto. From 2002 to 2008 he was the founding director of the Intellectual Property Research Institute of Australia (IPRIA), a national centre for multidisciplinary research on the economics, law and management of intellectual property, based at the University of Melbourne. Andrew has authored or edited more than one hundred publications in the field of intellectual property, with a particular focus on the application of copyright and trademark laws to the digital environment and on patent protection for biotechnological innovations. Andrew has been influential in the development of intellectual property law and policy, both nationally and internationally. He is a former member of the Australian government's Copyright Law Review Committee, and is a current member of its Advisory Council on Intellectual Property. He has produced commissioned studies for the Organisation for Economic Co-operation and Development, and been an Expert Consultant to the World Intellectual Property Organization. In 2005 he was identified by the international magazine *Managing IP* as one of the world's fifty most influential people in intellectual property.

Mark Dodgson is Professor and Director of the Technology and Innovation Management Centre at the University of Queensland Business School. He has produced eleven books and over one hundred academic articles and book chapters on innovation. Mark has advised numerous companies and governments throughout Europe, Asia and North and South America and has researched and taught innovation in over fifty countries. He is a founding director of the Think, Play, Do Group, an innovation advisory company. Mark's PhD is from Imperial College London. After eight years at the Science Policy Research Unit at Sussex University he was appointed Foundation Professor of Management at the Australian National University (1993–2002) where he was executive director of the National Graduate School of Management. He is a Fellow of the Academy of the Social Sciences in Australia, Fellow of the Royal Society of Arts, Fellow of the Australian Institute of Management and Visiting Professor at Imperial College London. He is editor-in-chief of *Innovation: Management, Policy and Practice*, a journal he founded in 1997, and a member of the editorial board of seven other academic journals, including *Research Policy*. Mark's current research includes studying the role of science in services innovation, innovation in cities, the consumption of innovation, dynamic social networks, innovation in complex projects, theory of innovation policy and the life of Josiah Wedgwood (1730–1795). His latest book (with David Gann) is *Innovation: A Very Short Introduction* (Oxford University Press, 2010).

Brian Fitzgerald studied law at the Queensland University of Technology (QUT) graduating as University Medalist in Law and holds postgraduate degrees in law from Oxford University and Harvard University. He is well known in the areas of intellectual property and Internet law and has worked closely with Australian governments on facilitating access to public-sector information. Brian is also a project leader and active member of the Creative Commons community. He was Head of the School of Law and Justice at Southern Cross University (1998–2002) and Head of the School of Law at QUT (2002–2007). Brian is currently a specialist Research Professor in Intellectual Property and Innovation at QUT and through his work with the ARC Centre of Excellence for Creative Industries and Innovation has been appointed an Honorary Professor at the City University of London. In 2009 Brian was also appointed to the Australia Government's "Government 2.0 Taskforce" and the Advisory Council on Intellectual Property (ACIP).

Joshua Gans is the foundation Professor of Management (Information Economics) at the Melbourne Business School, University of Melbourne. He has been at the Melbourne Business School since 1996. Prior to that he was at the School of Economics, University of New South Wales. Joshua holds a PhD from Stanford University and an honours degree in economics from the University of Queensland. At the Melbourne Business School he teaches MBA students in introductory microeconomics, technology strategy, economics of organisations, incentives and contracts and market design using game theory. He has also co-authored (with Stephen King and Robin Stonecash) the Australasian edition of Greg Mankiw's *Principles of Economics* (published by Cengage), *Core Economics for Managers* (Cengage), *Finishing the Job* (MUP) and *Parentonomics* (http://parentonomics.com/; New South/MIT Press). While Joshua's research interests are varied, he has developed specialties in the nature of technological competition and innovation, economic growth, publishing economics, industrial organisation and regulatory economics. This has culminated in publications in the *American Economic Review, Journal of Political Economy, RAND Journal of Economics, Journal of Economic Perspectives, Journal of Public Economics* and the *Journal of Regulatory Economics*. Joshua serves as co-editor of the *International Journal of Industrial Organization*, as associate editor of the *Journal of Industrial Economics* and is on the editorial boards of the *BE Journals of Economic Analysis and Policy, Economic Analysis and Policy, Games* and the *Review of Network Economics*. In 2007, Joshua was awarded the Economic Society of Australia's Young Economist Award (www.ecosoc.org.au/cc/awards). In 2008, Joshua was elected as a Fellow of the Academy of Social Sciences in Australia.

Leon Mann holds a BA and an MA in psychology from the University of Melbourne and a PhD in social psychology from Yale University. He has taught at Harvard University, University of Sydney and Flinders University. He is a past President of the Academy of the Social Sciences in Australia and of the Australian Psychological Society. From 1991 to 2003, he taught at Melbourne Business School where he was Pratt Family Chair of Leadership and Decision Making. Leon has held Australian Research Council Linkage Grants with leading Australian companies in the areas of R&D leadership, teams, knowledge sharing and effective collaboration in research organisations. He served on the Australian government's National Research Priorities Consultative

Panel and Expert Advisory Committee 2002–2003. He is currently director of the Research Leadership Unit in the Melbourne Research Office at the University of Melbourne and a Professorial Fellow in the Faculty of Medicine, Dentistry and Health Sciences. His publications on creativity and innovation have been on factors influencing innovation; models of leadership and communication; creativity in project teams; team climate and innovativeness; leadership and trust in project teams; and factors in successful research collaboration between industry and universities. He is the co-author with Irving Janis of *Decision-Making* (1977), and his other books include *Developing Leaders in R&D* (1994) and *Leadership, Management and Innovation in R&D Project Teams* (2005).

Jane Marceau, BA London School of Economics, PhD Cambridge, Fellow of the Academy of Social Sciences in Australia. Jane Marceau has taught and researched in the UK, France and Australia. Since retiring, she has been a Visiting Professor at the City Futures Research Centre, University of New South Wales. She is best known for her long-term work on innovation, most recently innovation in cities. She is a frequent consultant to governments at both state and federal level and has been a member of numerous government research and other policy advisory committees. She is the author or co-author/editor of six books, more than fifty papers and numerous research reports.

Erica McWilliam is an internationally recognised scholar in the field of pedagogy with a particular focus on workforce preparation of youth in post-compulsory schooling and in higher education. She is well known for her contribution to educational reform and its relationship to "over the horizon" work futures in the context of the new knowledge economy across the entire spectrum of formal learning environments from early years to doctoral education. Her career has involved four decades as an educator, moving from two decades in the schooling sector to a professorial role as an educational leader in the Queensland University of Technology, Australia. She has also worked as a professor of education in the Centre for Research in Pedagogy and Practice in the National Institute of Education in Singapore. Erica continues to play a leading role in the Creative Workforce 2.0 Research Program in the Australian Research Council's Centre of Excellence for Creative Industries and Innovation, led by QUT. Her research and scholarship continues to be well known for its focus on creative capacity building for twenty-first-century employability. Her latest sole-authored book, *The Creative Workforce: How to Launch Young People into High Flying Futures*, was published with UNSW Press in 2008. Erica recently accepted a part-time appointment as an Education Futurist at Brisbane Girls Grammar School, Queensland.

John Sweller is Emeritus Professor of Education at the University of New South Wales. John holds a BA (Hons) and a PhD from the University of Adelaide and has been a Fellow of the Academy of Social Sciences of Australia since 1993. His research is in the area of cognition and instruction. In the 1970s and early 1980s he was engaged in isolating some general principles influencing learning and problem-solving performance. Since then, he has been concerned with applying these principles to learning and problem-solving in most curriculum areas. With many colleagues, both local and international, John has developed cognitive load theory as part of this process. In recent years he has provided a theoretical justification for the suggestion that the information structures that underlie evolution by natural selection also underlie human cognitive architecture. All common citation indices record that John's work has been cited on many thousands of occasions with a rapidly increasing citation rate. For the last decade, one or two special journal issues devoted to cognitive load theory, an instructional theory that he helped originate, have appeared every year.

Simon Ville is Professor and Head of the School of Economics, University of Wollongong, primarily specialising in the teaching and research of the economic and business history of Britain, Europe, Australia and New Zealand. He is a Fellow of the Academy of Social Sciences in Australia and President of the Economic History Society of Australia and New Zealand. He was editor of *Australian Economic History Review* (1997–2003) and serves on the board of international journals and academic societies including *Business History Review* (US) and *Business History* (UK). He has written extensively for leading international journals in economics, economic and business history, management and law. His most recent books are *The Big End of Town: Big Business and Corporate Leadership in Twentieth-Century Australia*, with G. Fleming and D. Merrett (Cambridge University Press, 2004); *How Organisations Connect: Investing in Communication*, edited with Professors S. Macintyre and G. Boyce (Melbourne University Press, 2006). He has held a range of Australian Research Council (ARC) grants. His current ARC research grants deal with the economic and social consequences of Vietnam era conscription, the retention of skilled migrant women in Australia and industry associations and wool marketing in Australasia.

Jonathan West is Professor and founding director of the Australian Innovation Research Centre, hosted by the University of Tasmania. Prior to his current appointment, Professor West spent eighteen years at Harvard University, where he was Associate Professor in the Graduate School of Business Administration. His doctoral and master's degrees are from Harvard University, and he holds a Bachelor of Arts majoring in the history and philosophy of science from the University of Sydney. Jonathan's research focuses on understanding the roots of superior performance in innovation systems, particularly in the fields of agribusiness, the life sciences and biotechnology. He was formerly faculty director of the Harvard Life Sciences Project, a multi-faculty university initiative to understand the economic dynamics of the ongoing revolution in biology. He is the author of a detailed study of advanced technology development in US and Japanese semiconductor companies, and has studied the technology development strategies of leading electronics firms in the United States, Japan, Europe, Korea and Taiwan. Jonathan has served as a consultant to and board member of major corporations around the world and as an advisor to several governments, particularly in the fields of agribusiness and life sciences. His research has appeared in many scholarly journals and several books. Jonathan lives with his wife, Susan, and daughter, Eliane, on their family farm overlooking the Freycinet Peninsula on the East Coast of Tasmania.

Index

moral hazard 84
Morrell, E. 205
motivation: extrinsic 143; intrinsic 143, 155
multilevel statistical analysis 251
multiple discoveries 4
Mumford, M. 235
myths, about creativity and innovation 3–5, 223, 239

N
nanotechnology 246
narrow limits of change principle 230–1, 234
National Center on Education and the Economy (NCEE), US 212, 218
national innovation systems 9, 15–31, 51, 52, 73, 256, 258–60; components of 35–8, 46, 52; defined 16; development and evolution 16–18; innovative capacity 262; research facilities 17–18
natural information-processing systems 228–33; borrowing and reorganising principle 229–30; environmental organising and linking principle 231–2; information store principle 228–9; narrow limits of change principle 230–1, 234; randomness as genesis principle 230, 233–4, 236
Neale, M. 165
Neill, A. S. 209
Netherlands 67
networks: industry 247–8; *see also* social networks
new capital issues 75–6
new generation innovation model 34
new technologies 33–4, 35, 45, 46, 218, 245–6, 260; social impacts of 268
Newton, I. 92
North, D. C. 67
Northern Territory v. Collins (2008) 125
"not invented here" (NIH) syndrome 164, 173

O
OECD 60, 61, 255, 261, 262; OECD Reviews of Innovation Policy 258
open innovation 39, 40, 44, 58, 257, 265
open source innovation 264

Oppenheim, J. R. 268
organisational behaviour 171
organisational culture/climate 142, 156, 161–2, 167
Organisational Supports for Innovation Questionnaire (OSIQ) 162
organisations 11, 17, 265–6; centralisation 189, 194–5, 265; decentralisation 189, 194, 195–6, 265; information-processing systems 189, 191–6; project-based 195; research 265; responses to complexity and uncertainty 189, 194–6, 265
Orica company 158, 161
originality 110, 112
originators 66
Osborn, A. 212

P
packaging design 182–3
Palo Alto Research Center (Xerox PARC) 144, 155
Parnes, S. 212
Patent Lens 130
patent system 69, 85–6, 87–8, 91, 94, 129–31, 259, 264; databases 130; exclusive rights 107, 110, 111, 114; ingenuity threshold requirements 112; Peer to Patent project 130–1; secondary infringement 125–6
path dependency 66
pathways, creativity-innovation 242, 243–50, *see also* methods for studying 250–252
Patterson, F. 211
people factor in city innovation 56–7
perception 137. *see* seekers
Perkins, D. 215, 216
Perry-Smith, J. E. 142
personality traits and creativity 137
Peters, T. 156
pharmaceuticals industry 19, 20, 25, 26, 85, 100, 264
Piaget, J. 234
Picasso, P. 138
Pirola-Merlo, A. 160, 162, 166
place(s) 9, 50–63, 248–9, 261–2
play, and creativity 11, 170, 176–86, 244, 246, 265
Plucker, J. 224
policy implications and recommendations 258–68; city innovation